I0464039

DEBUNKING DARWIN

Natural Selection Is NOT *Science*

JOSEPH
ANDERSON

Copyright © 2014 Joseph Anderson
All rights reserved.
ISBN: 1482612984
ISBN 13: 9781482612981

I dedicate this book to:

my mother and father, who gave me love and a good start in life;
my three brothers and sister, who gave me much help along the way;
Mrs. O'Brien, my fourth and fifth grade teacher in
Public School Number 2, Jersey City, New
Jersey, who encouraged me and provided me with a
strong foundation that served me all my life;
my wife, Jane, a gifted teacher who gave me all that is meaningful in life: an agreement to share our lives,
a warm home, and three great children. They
gave us new ways to see the world.

PREFACE

Charles Darwin is well-known for the theories of evolution by natural selection and survival of the fittest. He is less known for his theories of sexual selection, the tree of life, common ancestors (fossils and living creatures descending from one simple life form), humans' descent from apes, and the role of God. This book will show Darwin's theory of evolution as gravely flawed, like many of his other ideas—so much so that it must operate through testimonies outside of nature and science, thus operating like models of major religions. This book explores Darwin's theory for what it is and is not, unable to meet even the simplest demands of nature or science. While Darwin himself used the term *evolution* correctly (as an effect, "that which was created"), this book will clearly define evolution and demonstrate today's frequent misrepresentation as a process of creation rather than something that is created.

I am a licensed professional engineer (retired). In 1991, I began reading about natural selection in a single book about evolution by Jeremy Rifkin, titled *Algeny*. That event began years of reading about evolution by natural selection, peeling back layer after layer in the process. As I learned, natural selection has a direct relationship to Christianity, Judaism, Islam, and many other faith systems. Though not immediately apparent, it even has a direct relationship to the early concerns of practicing religion freely in the newly forming country that would become the United States. At that time, Christians feared that the government would impose a state religion on them as in England, where one Christian group imposed its religion on other Christians. Because of that concern, the Danbury Baptist Association sent a letter to then-president Thomas Jefferson in October 1801, which contained the following concerns:

Our sentiments are uniformly on the side of religious lib-
erty—that religion is at all times and places a matter between
God and individuals—that no man ought to suffer in name,
person, or effect on account of his religious opinions—that
the legitimate power of civil government extends no further
than to punish the man who works ill to his neighbors.[1]

Jefferson responded on January 1, 1802. His response included an excerpt
from his famous "wall of separation between church and state," as follows:

Believing with you that religion is a matter which lies solely
between man and his God, that he owes account to none other
for his faith or his worship, that the legitimate powers of government reach actions only, and not opinions, I contemplate
with sovereign reverence that act of the whole American people which declared that their legislature should make no law
respecting an establishment of religion, or prohibiting the free
exercise thereof, thus building a *wall of separation between
church and state.* [Italics added][2]

The phrase "wall of separation between church and state" does not appear
in the US Constitution[3] but rather in the principle of freedom of religion,
which is one of five freedoms stated in the First Amendment:

Congress shall make no law respecting an establishment of
religion, or prohibiting the free exercise thereof.

The phrase "wall of separation between church and state" was written in a
letter by Thomas Jefferson and meant to show that the government would not
adopt an official state religion to be imposed on the country. It was meant to
show that one faith system would *not* be imposed on all people, as was done in
England and other countries. However, a religion is now being imposed in the

United States: the religion of Darwinism. It is imposed in science and biology textbooks.

When a person is aware that natural selection is not part of nature and is not science, he or she can more easily see the beliefs inherent in it being imposed on people using a label of "science" through textbooks and in classrooms. It is imposed as a way to "think about" and "speak about" nature. In other words, the religion of Darwinism is being made an official state religion. The debate about evolution is not one of science against religion, but one of religion against religion: the religion of Darwinism against any competing religion, such as Christianity, the largest practicing religion in the United States.

Joseph Anderson

CONTENTS

CHAPTER 1

EVOLUTION DEFINED

Natural Selection is not Evolution. Yet, ever since the two words have been in common use, the theory of Natural Selection has been employed as a convenient abbreviation for the theory of Evolution by means of Natural Selection, put forward by Darwin and Wallace. This has had the unfortunate consequence that the theory of Natural Selection itself has scarcely ever, if ever, received separate consideration.[4]

—R. A. Fisher, 1930

Sir Ronald Aylmer Fisher (1890–1962) was an English statistician, evolutionary biologist, geneticist, and eugenicist. Fisher is known as one of the chief architects of the neo-Darwinian synthesis. Anders Hald[5] called him "a genius who almost single-handedly created the foundations for modern statistical science," and Richard Dawkins named him "the greatest biologist since Darwin."[6]

Evolution: Introduction

Fisher's above statement that "natural selection is not evolution" emphasizes natural selection and evolution are two separate and non-interchangeable parts of one "cause and effect" creation model. Others in Darwin's time, such as Thomas Huxley, Samuel Butler, and Alfred Russel Wallace (the co-discoverer of natural selection), showed the two parts of Darwin's "cause and effect"

1

creation model were not interchangeable; each is quoted later with Wallace quoted extensively in chapter 11, "Testing Miracles and Natural Selection." Darwin correctly used "cause and effect" by proposing "evolution through natural selection," Darwin even argued against another cause of evolution, God, which is discussed later. Wallace, the "co-discoverer" of natural selection, wrote the same about natural selection being separate from evolution. For him, God was the cause of evolution in some cases and natural selection was the cause evolution in other cases. The usage of evolution as a cause is misleading, incorrect, and confusing. Evolution and natural selection cannot be interchanged, and the terms cannot serve as abbreviations for one another without sacrificing the model's consistency, credibility, and legitimacy.

The first man in England to propose naturalistic evolution was Erasmus Darwin, Charles Darwin's grandfather. In 1794, Erasmus published the first volume of a work titled *Zoonomia*, followed in 1796 by a second volume.[7] In Volume I of *Zoonomia*, Erasmus published his ideas on evolution and its mechanisms.[8] This sold well; became well known; and was even translated into German, French, and Italian, making Erasmus Darwin famous. Erasmus Darwin preceded his grandson in many evolutionary concepts, which Charles would later use in his 1859 book, *On the Origin of Species By Means Of Natural Selection, or the Preservation of Favoured Races in the Struggle for Life*, deleting the word "On" from the title in the sixth edition. Charles Darwin was born with the advantage of being the grandson of a famous doctor and author.

Evolution only makes up half of the picture of creation: it shows the picture of what was created, namely the fossils and living creatures. The cause of creation forms the other half of the picture. That half cannot be found in the earth. It cannot be found among excavated fossils. Missing from evolution and missing from all the fossils, whether already excavated or yet to be discovered, is the mechanism of their creation: the *cause*. That cause, together with the effect, makes up the model of how "cause and effect" link together. It consists of the parts of nature that operate to cause creation. Those parts of nature need to be named and described, together with their processes and the rules and relationships that govern them during the creation process. Only a creation model shows all of nature responsible for the fossils' creation; only that creation model properly demonstrates cause and effect. Without the model of cause and effect in nature, all we have is testimony, or people telling others what they think and believe causes evolution. Without a creation model that shows nature's operations, evolution is argued from inferences and interpretations of the past that some call "historical." When historical inferences are used for cause and effect, the model merely remains on the sidelines, a point of reference and attribution during the testimony.

Evolution does not change with each different cause that is claimed; it remains the same, no matter what cause is said to be responsible for its creation. People may champion any of the creation models and over time have done just that, with the fossils remaining the same. Today, natural selection is supported as a cause as it has always been supported—testimonially. Past causes of evolution have included Lamarck's "use and disuse," special creation in Genesis (the first book of the Hebrew Bible and the Christian Old Testament), evolutionary arms, race, orthogenesis, and God. Different groups supported different causes of creation, but what was created always remained the same. Different views, arguments, interpretations, and portrayals of those fossils were raised, such as those favoring natural selection, orthogenesis, use and disuse, and God. For example, Charles Lyell, Thomas Huxley and Alfred Russel Wallace (in some cases) objected to natural selection being the creation model. The fossils' creation is called evolution and remains the same for natural selection, God, or any other cause. Evolution never changes, even if the cause is thought to change. The creation of new creatures is what causes evolution over time. Something caused the creation of each of the new creatures that shows the sequence of creation in time. That something is called a creation model. It is the cause of creation that is argued in the public debate about evolution.

The Swiss philosopher Bonnet (1720–1793) coined the term "evolution" and first used it for the preformation theory of individual development.[9] He coined it long before Darwin published his 1859 *Origin of Species*. However, Bonnet's usage does not match the modern use of the term. Evolution has had different meanings over time, but in no case has evolution ever constituted a process of creation—or any process at all—or been a synonym for natural selection or descent with modification. In all cases, evolution stands apart from its cause, which is the cause of creation.

Evolution is not simply a matter of change over time. Change may take place by tree leaves growing and falling; tides rising and ebbing; day turning into night; and offspring having varying sizes of heights, arms, legs, noses, weights, and other body part differences: all of this is change, and none of it is evolution. "Change" is a very broad brush that incorrectly paints too much as being evolution. It may be suitable to describe the weather, but not evolution. Evolution is a particular type of effect that shows appearances of new body parts, new appendages, and new fully completed creatures. When organized into a table of creatures, it becomes the "table of evolution": the table shows all fossil creatures, but it does not show any cause of their creation. The "table of evolution" shows all that was created, not someone's interpretations. The fossil creatures in the table of evolution are not changing. But the table is changing with each new creature that appears in it. Those changes show evolution, but not the cause of their creation.

The eighteenth-century embryologist[10] Charles Bonnet used the term "evolution" to describe the adult form of the embryo's development. He applied this usage to evolution before the modern usage of the word. At that time, the term "evolution" showed a gradual forming of a new person in the womb of the mother. The embryo was caused, and its creation is the effect. Applied to embryos, the beginning and end of evolution were known. They constituted the embedded "plan" that followed a genetic coding. The end result was the creation and birth of an existing form of a creature, different from its parents only by degrees, with differences appearing as normally distributed variations of the same identical body plan. For example, with humans, the body plan contains a skeletal system, two feet, two hands, one head, and ten major systems of organs. This definition of evolution added no new creatures, for it was "like begets like." In this way, the term "evolution" made sense and was correct. Embryos still have a known beginning (a plan embedded in a genetic code) and a known end (the birth of a creature like its parents).

Researchers knew little about genetics at that time. The "e" in evolution means "out," coupled with "*volutio*," which means "turning or folding." Thus, evolution means an "unfolding" or an "unrolling"[11] (as with unrolling a scroll to discover what lies inside). Darwin rarely used the term "evolution," although it is now so commonly associated with his creation model of natural selection. In Darwin's time, evolution usually indicated some sort of progressive process of creation, like orthogenesis, and had been commonly used since at least 1647.[12]

The early evolutionary thinkers of the eighteenth and early nineteenth centuries had to invent terms to discuss their ideas, and their terminology did not stabilize until sometime after the publication of the first edition of Darwin's *Origin of Species* in 1859. Peter J. Bowler wrote an excellent twenty-page article on the changing meaning of evolution.[13] The word "evolution" was quite a latecomer. It can be seen in Herbert Spencer's 1851 work *Social Statics,* where Spencer did not use the term as part of new creatures being formed. Two examples of Spencer's use of the term "evolution" are as follows:

> In consequence of the probability, or perhaps we may say the certainty, that the causes leading to the evolution of a new idea in our mind, will eventually produce a like result in some other mind, the claim above set forth must not be admitted without limitation.[14]

> But note lastly, and note chiefly, as being the fact to which the foregoing sketch is introductory, that what we call the moral law—the law of equal freedom, is the law under which individuation becomes perfect; and that ability to recognise and act up to this law, is the final endowment of humanity—an endowment now in process of evolution.[15]

The term "evolution" was not in general use until about 1865 to 1870.[16] Darwin used it for the first time in 1871, on the second page of the first edition of his book, *The Descent of Man.* He writes:

Of the older and honoured chiefs in natural science, many unfortunately are still opposed to evolution in every form.[17]

In the above quote, Darwin referenced "naturalistic" evolution, hence the opposition to it. Prior to Darwin and to the use of the term "evolution," God was recognized as the creator of all creatures and man, as shown in the first book of the Bible, Genesis. Knowing this, it is easy to see that "evolution by God" was taking place. In the quote that follows, Darwin shows he understands that evolution is an effect, and he proposed natural selection as the cause: he used the phrase "evolution through natural selection."[18] Darwin did not use the term "evolution" in his publications of *Origin of Species* in 1859, 1860, 1861, 1866, or 1869. It was not until the sixth and final publication of *Origin of Species* in 1872 that Darwin used the term eight times. As an example, he writes:

If numerous species, belonging to the same genera or families, have really started into life at once the fact would be fatal to the theory of evolution through natural selection.[19]

In *Origin of Species*, Darwin waged a war against evolution by special creation, which is evolution by God, proposing instead evolution by natural selection. He claimed that evolution was caused by natural selection, showing evolution to be an effect. Some interpret evolution as "descent with modification through natural selection."[20] Used in this way, "descent with modification" represents variation, and "through natural selection" means "selection and accumulation in a direction." The complete phrase "descent with modification through natural selection" merely describes natural selection's two parts using different words. However, the simple phrase, "descent with modification," if left unqualified, has nothing to do with evolution. It is merely *stasis*, meaning no new creatures are being created. When descent with modification operates through a male-female parental recombination of genes, it results in an unchanging, common body plan. When no mutations of the genetic material take place, nothing new is theoretically possible in nature.

Darwin's creation theory eventually waned, but was revived with the modern synthesis and the publication of many texts by different authors. In that synthesis, natural selection uses genetic mutations for gradual evolution to take place. That claim did not appear until sometime between the 1930s and 1950s. The modern synthesis was, in essence, the modern series of arguments by many authors supporting natural selection and against special creation (David Lack, George Gaylord Simpson, Julian Huxley, and Thaddeus Dobzhansky, among others). Darwin writes repeatedly that evolution is produced by a cause of "some form" in the 1872 sixth and last edition of *Origin of Species*:

> At the present day almost all naturalists admit evolution under some form. Mr. Mivart believes that species change through "an internal force or tendency," about which it is not pretended that anything is known.[21]

For Darwin, the creatures created are the *effects*, meaning the creatures were caused by something. In this case, that something was an "unknown internal force" or tendency called "orthogenesis." For Darwin, creatures were created in one of two ways: by God or by nature (meaning natural selection). Darwin denied that God created creatures directly through special creation or independent creation. He writes to Lyell on 12 March 1863:

> Either species have been independently created, or they have descended from other species, like varieties from one species.[22]

In Darwin's view, creation occurred directly by God, which he argued against, or by natural selection, which he believed was the case. In either case, evolution takes place when new creatures are created, with only the cause being different. It is curious that Darwin thought that there are only two options for creation of new creatures: an object of religious faith (God) or natural selection, which he calls "science" and which he claims is operating in nature. In some people's minds, if you argue God away, then natural selection is the only answer. This is not how a scientific causal model is determined.

The Rate of Evolution: Gradual, Punctuated, or Instantaneous Creation (Saltation)

Darwin denied the abrupt evolutionary changes by saltation and sudden large-scale mutation[23] as he thought that the creation of a new creature could not take place in one generation or instantly. His claim was not a scientific statement but rather something he believed in. Despite the fact that all he observed around him were fully completed creatures with all the characteristics of saltation and no intermediate creatures that were transitioning variation by variation, he held that the intermediate creatures were there at one time but are now missing. A gradual creation model leaves incompletely formed body parts and creatures existing in intermediate stages of variation by variation accumulation. His model of natural selection has no reference to time, but if it did, it could otherwise show that saltation did or did not take place. Without time in the model, Darwin had only his arguments that saltation did not take place. Others in his day (such as Samuel Butler and Thomas Huxley) disagreed with him about saltation and thought that such creation events did take place.

The fact that time is not contained in any model of creation makes it theoretically impossible to determine a rate of creation or evolution. One has to be satisfied with existing effects, such as radioactive dating, and attribute them to the model of creation. Evolutionary rates are difficult to measure accurately and impossible to know theoretically, which necessitates the use of terms such as "gradual," "rapid," "punctuated," "instantaneous" (saltation), and "one generation," which commonly appear in evolutionary literature with no theoretical relation to the models of creation. One may find "evolution" taking place in one day or one million years.

Gradual creation may be called "phyletic gradualism," as Stephen J. Gould turned a phrase for it. Natural selection offers no indication of creation as gradual, instantaneous, or otherwise, as it does not contain any means of relating to time or the creation rules of nature. In addition, the adjectives "gradual," "punctuated," "saltation," and others work independently of nature, representing schools of thought, not acts of nature as captured in a model of creation. Niles Eldredge and Stephen J. Gould wrote the 1972 paper titled "Punctuated Equilibria: An Alternative to Phyletic Gradualism," which conflicted with

Darwin's gradual creation model. The two authors proposed "punctuated equilibrium" as a way to think about fossil evolution using a rapid or sudden creation where creatures appeared in a punctuated manner. After punctuated creation, the creatures remained the same, in an equilibrium that was stasis, meaning they were not changing into new creatures but rather were varying about a common body plan. Both Eldredge and Gould were opposed to special creation:

> The history of life is more adequately represented by a picture of "punctuated equilibria" than by the notion of phyletic gradualism. The history of evolution is not one of stately unfolding, but a story of homeostatic equilibria, disturbed only "rarely" (i.e., rather often in the fullness of time by rapid and episodic events of speciation).[24]

In other words, new creatures appear suddenly and remain the same. That is, they remain in stasis, unchanging, varying about a common body plan, and then they disappear from the fossils. No gradual change occurs over time as claimed by Darwin and supporters of natural selection. The coelacanth, lungfish, horseshoe crab, and other living fossils provide examples of creatures that do not change over time. Another example is the lack of evolutionary process in the origin of frogs. The oldest known frogs are completely different from fish, first appeared with their own unique structures, and possessed exactly the same characteristics as modern frogs. There is no difference between the approximately twenty-five-million-year-old fossil frog in Dominican amber and living specimens.[25] Stasis is the rule. It was in Darwin's time and is today as well.

Whether one thinks God or natural selection created all of the creatures, or that creatures arose gradually or by saltations, the creatures in the table of evolution remain unchanged and an accurate representation of the creatures as they were created. Only fully formed creatures appear in that table of evolution for good reason: that is all there is. Nothing is missing. Nothing is incomplete in its creation, as would be the case for gradually created body parts and bodies. Even creatures claimed to be intermediates appear fully completed, meaning one fully formed creature following another with no partially completed ones in between.

Butler in 1882: Evolution Is Separate from Natural Selection

The meaning of evolution stays the same, with or without a known cause; it is each creature in the sequence of its appearance on earth. This holds true even if we don't know how they were created. Samuel Butler, in his 1882 book *Evolution Old and New*, clearly tells us that evolution is not a cause and not a process. That was true then, and it is still true today, over a hundred years later. Butler addressed natural selection and survival of the fittest not as being cause-and-effect theories but as personal descriptions. In 1882, Butler writes:

> Independently of the fact that "natural selection," or "the survival of the fittest," is in no sense a theory, but simply an observed fact, yet even if the words are allowed to stand for "descent with modification by means of natural selection," it is still misleading to write as though this were synonymous with "the theory of evolution," or "the theory of descent with modification."[26]

Survival of the fittest means that creatures survive in differing circumstances, and that the ones who live are the "fittest," with the key terms ("fittest," "selection," "modification") left undefined in nature; this amounts to a description, not a cause-and-effect model of evolution. It literally tells us nothing about nature or creatures except that they are either alive or not. The same holds true with natural selection, which merely describes living creatures as "naturally selected." Those that have died were not selected. Butler separates cause from effect clearly in the following statement:

> To present evolution as a "process" or "cause" prevents the reader from bearing in mind that "evolution by means of the circumstance-suiting power of plants and animals [fitness]"

as advanced by the earlier evolutionists; and "evolution by means of lucky accidents" with comparatively little circumstance-suiting power, are two very different things, of which the one may be true [evolution] and the other untrue [the cause of evolution].[27]

Some definitions of evolution wrongly embed natural selection as a "fact of evolution," namely that "all organisms living and dead are the end products of a natural process of development from a few forms, perhaps ultimately from inorganic materials ("common descent")."[28] As Butler describes, evolution is not a process of creation or a cause of new creatures. Butler then shows how mixing, confusing, and equivocating the terms used in support of Darwin's evolution by natural selection misleads readers. Butler writes:

The mixing of evolution's "effect" and "cause" leads the reader to forget that *evolution by no means* stands or falls with *evolution by means of natural selection*, and makes him think that if he accepts evolution at all, he is bound to Mr. Darwin's view of it.[29]

Many then, as today, associate evolution with Darwin, which is incorrect. Evolution is part of nature, the effect. Darwin did not propose evolution. He proposed the cause of evolution when he claimed "evolution through natural selection,"[30] showing evolution being caused by natural selection. The separation of natural selection (cause) and fossils (effect) by Darwin shows he understood cause and effect as two separate parts of a model of creation. Scientists understood this in Darwin's day but have apparently forgotten it today, or at least the supporters of natural selection have forgotten it. It is important to know, as Butler tells us, that one can accept evolution and not accept a claimed cause. Darwin did just that when he rejected special creation: he rejected a cause but kept evolution. That is, one can accept the effect and not accept what others claim to create or cause the effect. Butler certainly did just that when he denied God as the cause of evolution, or simply "evolution by God." Butler accepted evolution but denied God's role in creating it, which is a theoretically legitimate position. The same can be done for natural selection, where one can accept evolution and reject natural selection. Many have done just that, including Darwin's friend, Thomas Huxley. If one rejects natural selection, then that rejection also applies to "descent with modification through natural selection"[31] as well. Modifications are merely physically undefined and unknown variations claimed by arbitrary attribution. They are not fact.

Huxley: Evolution Is Separate from Natural Selection
Similar to Butler in the above quote, George Mivart writes that Thomas Huxley said much the same thing: a person can accept evolution and its claimed cause or can even reject all of the claimed causes and still accept evolution. Evolution stands completely independent of any cause. Huxley writes:

> I can testify, from personal experience, it is possible to have a complete faith in general doctrine of evolution [the effect], and yet to hesitate in accepting the Nebular, or the Uniformitarian, or the Darwinian hypothesis [as the cause] in all their integrity and fullness.[32]

Huxley recognizes the separation of cause from effect with evolution. The point remains that evolution is not a process; it is not "descent with modification." Evolution is created by a creation model, and that model alone, when successful, is the cause. In 1871, one year before Darwin's sixth and final publication of *Origin of Species*, George Mivart comments on Thomas Huxley's view of evolution being separate from some particular cause:

> It is quite consistent, then, in the professor [Thomas Huxley] to explain the difficulty as he does; but it would not be similarly so with the absolute and pure Darwinian.[33]

Darwinists appear to accept, to this day, that evolution is a process that creates new creatures—or at least they portray it as such a process. The equating

12

of evolution and natural selection, evolution and survival of the fittest, or evolution and descent with modification mixes the terms, perhaps to lead others to believe that the term *evolution* represents natural selection, which is incorrect and misleading. One may make the case for Darwinists being so bound to their "no special creation" worldview that they can only observe evolution as inextricably linked to *their* creation model. This may serve as an example of "fundamentalist faith," albeit secular faith.

The First Cells Start Evolution: Beginning the Table of Evolution

The appearance of the first living creatures as fossils shows a unique kind of evolution, a change from an absence of creatures on earth to the creation of the first completely formed living cells. These first fully completed creatures were formed about 3.5 billion years ago, 1.2 billion years after the earth formed.[34] These cells became the first entry into the table of evolution, the table that lists all living creatures in their order of appearance on earth. Afterward, for 2.5 billion years, only these fully formed, single-celled prokaryotes existed.[35] No other known creatures existed on earth during that time. These first living creatures ranged in size from four-millionths to one ten-thousandth of an inch in diameter, which is 0.0001 mm to 0.003 mm or 0.000004 inches to 0.0001 inches. With the exception of a few species, prokaryote cells are surrounded by a protective cell wall.[36]

From nothing at all, the cell's material, shape and dimensions, parts locations, waste removal, food distribution and other sustainment processes, and operations of internal systems (such as control-feedback mechanisms) made up the many parts of the cell that had to be created, for they did not exist at the beginning of the earth's creation. The cell wall and all its internal parts were created from those components and systems. These creatures are not simple, as many internal and external systems functioned in an integrated, organized way to keep them alive. These early single cells contained the following parts:[37] bacterial flagellum (whip-like structure), capsule, cell wall, plasma membrane, cytoplasm, ribosomes, plasmid, pili, and nucleoid (circular DNA).[38] These parts worked together, interoperating as a team, giving life to the simple creature, a factory of life created from no life at all. Even the individual parts of the cell are not shown to exist prior to the first cells making one wonder, "How were these parts created, integrated, and made to operate?"

Evolution: Then Came the Eukaryote Cells, the Second Type of Creature

About 1.4 billion years ago, a second type of single-celled creature appeared fully formed: the single-celled eukaryotes.[39] The eukaryotic cells were much larger and more complex[40] than their predecessor prokaryotes. There exist no fewer than thirty-six phyla (different body plans) of mostly unicellular eukaryotes.[41] Eukaryotes include bacteria, blue-green algae, amoeba, and paramecium of high school biology laboratories,[42] as well as such utterly diverse organisms as radiolarians, foraminifera, sporozoans, zooflagellates, ciliates, green algae, brown algae, dinoflagellates, diatoms, Euglena, slim molds, and chytridiomycota.[43]

Evolution: Next Came the Multicellular Creatures

About seven hundred million years ago, the first multicellular life was created. No more changes to the population of creatures that lived on the earth appeared until the Cambrian explosion, which took place about 540 million years ago.

Evolution: The Cambrian Explosion of Life

After the first multicellular life existed without change, in stasis, for about 700 million years, many massive numbers of new creations took place in a short span of 5 or so million years, about 540 million years ago. Seemingly from nothing, in a geologic instant, the Cambrian explosion took place. Large numbers of assorted animal life were created in the geologic blink of an eye. As Stephen J. Gould tells us, in this "instant" of creation, enormous numbers of new, fully formed creatures appeared. Every phylum (classification) but one was created rapidly. All the "apparent" or "actual" designs of creatures appeared with functioning organs, major body systems, sustainment systems, maintenance systems, operating systems, and reproduction systems. Stephen J. Gould writes:

14

[S]tarting about 530 million years ago, constitutes the famous Cambrian explosion, during which all but one modern phylum of animal made a first appearance in the fossil record. (Geologists had previously allowed up to 40 million years for this event, but an elegant study, published in 1993, clearly restricts this period of phyletic flowering to a mere five million years.)[44]

The period of time before the Cambrian period can offer no comfort to Darwin's expectation of life's gradual accumulation.[45] The appearance of the newly created creatures, followed by stasis (long periods of no change in their forms), with no gradual appearances of new organs, new body parts, new feet or legs, or other appendages, does not bode well for a gradual accumulation of variations by natural selection. We see sudden creation and stasis. What we do not see is the cause of that sudden creation over approximately five million years.

Darwin's Missing Intermediate Fossils

When a new creature is created gradually, it necessarily follows that the entire body is not created at once, for it is formed variation by variation, accumulation by accumulation. Neither is each organ or system of organs created all at once but rather slowly, by variations. According to natural selection and necessarily to each gradual creation model, each variation is added to other variations that were created in prior generational years. In this gradual accumulation, the new body—whether a mouse, an elephant, or a whale—is formed by the accumulation of Darwin's "small" variations. These variations and accumulations were never defined in the model, which is the only place where science exists—if it exists there at all. The accumulation of variations and body parts into systems necessitates that each body part remain incomplete for many generations, all the while being inoperative. How many generations the body remains inoperative is unclear, perhaps thousands or millions. If creation were truly gradual, then the fossils would reveal those incomplete creatures. But intermediates fossils are not shown as incomplete creatures.

They are shown only as fully formed creatures. Incomplete intermediates don't exist in nature or in natural selection. Even the skeletons are fully formed and architecturally complete where only bones exist.

If all exhibited intermediates are fully completed creatures, then creation occurs by saltation, not by gradual, accumulated variations. Richard Milner, author of *The Encyclopedia of Evolution,* writes that there *are* intermediates. But he does not distinguish between the two types of intermediates that may exist: incomplete intermediates created gradually or saltation intermediates, created all at once and catalogued as intermediates. Using the fully formed saltation-type creatures, Milner writes of intermediates as if they were the partially formed, incomplete intermediate fossils Darwin said were missing:

> The oft-repeated claim that there are no transitional forms is demonstrably false. The Karroo region of South Africa, for instance, is a vast graveyard of the remains of mammal-like reptiles, a whole array of species whose [fully completed] anatomy was intermediate[46] between reptiles and mammals.[47]

> There is the famous *Archaeopteryx*, with its [fully completed] feathers, teeth, claws, and lizard-like skeleton, a transition between [fully formed] reptiles and birds.[48]

Milner published an excellent encyclopedia of evolution; and while likely not intending to do so, he made a good case for the instantaneous creation that is saltation, not gradual creation. The term "intermediate" was being equivocated, mixed up between fully formed creatures appearing in a museum catalog and partially formed creatures that were incrementally being changed into new creatures over many thousands of years: the two things are exact opposites and it is difficult to mix them up. The terms "transitional forms" or "intermediate forms" mean only one thing in terms of gradual creation: the fossil creatures exist with new, incomplete organs and new, incomplete body parts created by the accumulation of beneficial variations in a direction toward full completion. Those "transitional creatures" or "intermediates" are never shown to take place. What was missing for Darwin was never there, as observations showed in his time and still show to this day. Gradually created intermediates must be incomplete creatures with incomplete bodies, regardless of the cause of gradual creation. Richard Milner continues:

> And the African hominid fossils represent creatures with human-like dental patterns, small brains, arms longer than humans but shorter than modern apes, with pelvis, feet, and legs for upright walking.[49]

> Another evidence of transition is found in geographical distribution of living species. On Pacific island chains, for instance, biologists have tracked population species across thousands of miles discovering [fully completed] intermediate forms from one end of the island chains to the other.[50]

Everything that Milner mentions is fully formed. Fully formed creatures cited as intermediates are always saltations—created all at once. Darwin presented his beliefs about intermediate creatures in his 1859 *Origin of Species*. He referred to the intermediate creatures as transitional forms, which he did not define except by implication when he wrote that they were missing. Gaps, when mentioned, constitute the incomplete creatures that are never shown to exist but are said to be missing. Gaps constitute the incomplete creatures between fully completed ones. Darwin writes about the missing creatures as follows:

> [W]hy, if species have descended from other species by insensibly fine gradations, do we not everywhere see innumerable transitional [incomplete] forms? Why is not all nature in confusion instead of the species being, as we see them, well defined [and completed]?[51]

> But, as by this theory [of gradual creation by natural selection] innumerable transitional forms must have existed, why do we not find them embedded in countless numbers in the crust of the earth? It will be more convenient to discuss this question in the chapter on the Imperfection of the Geological Record; and I will here only state that I believe the answer mainly lies in the record being incomparably less perfect than is generally supposed. The crust of the earth is a vast museum; but the natural collections have been imperfectly made, and only at long intervals of time.[52]

Darwin may have overlooked the fact that what his gradual creation theory demanded was missing because his theory was wrong and the fossils were right. By claiming that the intermediary fossils were missing, Darwin admitted that neither he nor anyone else had ever observed any incomplete transitional forms when they should exist "in countless numbers," including slowly forming skeletal parts and many incoherent dead-end skeletal arrangements. That is, no fossil is shown to start as one fully completed creature, have its numerous offspring accumulate *good* variations to form every new body part gradually, and accumulate them into systems that form a new, fully completed creature. Darwin tells us that the fossil record is "imperfect." But the word "imperfect" reflects his expectations based on his worldview of gradual creation and a model that contains no parts of nature. If one were to use sudden creation or saltation, then the fossil record looks just as one would expect. There are no missing fossils, which is the only position that observations support.

If such a process as natural selection's gradual creation of new creatures ever took place, then untold millions of fossils should show the various stages of incomplete creatures that necessarily had to exist between two fully formed creatures: the original parents, then a series of incomplete intermediates, then the fully formed new creatures. Darwin's "descent with modification through natural selection"[53] necessitates that new bodies form through "selected" good variations that formed accumulations that became each new organ, adaptation, body part, and complete architecture. The modifications in "descent through modification" are the parts that are incomplete, the variations that are accumulating. A gradually created new creature would be 1 percent formed, then 25 percent formed, then 80 percent, and finally 100 percent over millions of years, and the fossils would show this scenario. Wondering why the fossils did not appear as he claimed for his model, Darwin writes:

> Why then is not every geological formation and every stratum full of such intermediate links? Geology assuredly does not reveal any such finely graduated organic chain; and this, perhaps, is the most obvious and gravest objection which can be urged against my theory. The explanation lies, as I believe, in the extreme imperfection of the geological record.[54]

> The abrupt manner in which whole groups of species suddenly appear in certain formations, has been urged…as a fatal objection to the *belief* in the transmutation of species. [Italics added][55]

> If numerous species, belonging to the same genera or families, have really started into life all at once, the fact would be fatal to the theory of descent with slow modification through natural selection…we continually overrate the perfection of the geological record, and falsely infer, because certain genera or families have not been found beneath a certain stage, that they did not exist before that stage.[56]

Why does Darwin hold his inferences as correct and others' opposing inferences as incorrect? Darwin's last quote stems from faith, not facts. Although the data is against him and others disagree with him, he maintains his belief in his model being correct. This is faith, but it is not science. The fossils show, again and again, that the data in the form of intermediate forms is not missing. It is there, but it is denied. The data shown by Darwinists as evidence for intermediate fossils are fully completed fossils, including their fully formed skeletons, which prove saltation, not gradual creation. Darwin continued with his defensive plea that his model of natural selection was correct:

> We continually forget how large the world is, compared with the area over which our geological formations have been carefully examined; we forget that groups of species may else-where have long existed…We do not make due allowance for the enormous intervals of time, which have elapsed between our consecutive formations,—longer perhaps in some cases than the time required for the accumulation of each for-mation. These intervals will have given time for the multi-plication of species from some one parent-form: and in the succeeding formation, *such groups or species will appear as if suddenly created.* [Italics added][57]

This is Darwin's "argument of hope"—that his model will be proven cor-rect in time. Supportive data that does not exist in nature exists as spirits in the minds for those believing in gradual creation and natural selection. These mental constructs are necessary to meet the expectations of a Darwinian worldview, which is not satisfied by the actual fossil record. Such expectations exist by faith, not nature—and not science.

Evolution Is a Fact, but "Evolution by Natural Selection" Is Not

Ernst Mayr, the late professor emeritus in the Museum of Comparative Zoology at Harvard University, writes that evolution is a fact.[58] Separately, in *The Encyclopedia of Evolution*, Richard Milner also writes that evolution is a fact.[59] The National Academy of Sciences (NAS) claims evolution is a fact.[60] There is no doubt that evolution is a fact of nature because evolution is the fossils that were created. However, evolution by natural selection is not a fact; it is a belief about the cause of the fossils creation. Looking at the fossils is looking at the "effect" called evolution. Placing the fossils in a particular

20

arrangement does not change that fact. They are observed so that the formation of evolutionary ideas about their causes can be made; they form evolution even though their cause of creation is unknown and being debated. Evolution is not creatures changing one into another for that is not a fact: it is a belief held by supporters of gradual creation and natural selection. Discard the cause of evolution and evolution still remains the same. Evolution is proven by the creatures that have been created and are observed as being fully formed. How we arrange the fossils and then attribute them to a cause is not a fact, but a testimonial assertion: a faith based claim. There are no facts scientifically linking fossils to their creation model, and only faith based testimony links the two—for every creation model that has been proposed. In his 2001 book *What Evolution Is*, Mayr writes:

> Evolution is not merely an idea, or a concept, but is the name of a process in nature, the occurrence of which can be documented by mountains of evidence that nobody has been able to refute...It is now actually misleading to refer to evolution as a theory, considering the massive evidence that has been discovered over the last 140 years documenting its existence. Evolution is no longer a theory, it is simply a fact.[61]

Mayr's assertion is incorrect. What Mayr does not mention is that his statement treats evolution and natural selection as if they were the same thing, which they are not. Evolution is not a process; it is only one-half of the process—it is the half that was created, the fossils, but it is not the cause of creation. Evolution is not responsible for creation. If it were, then natural selection would not be needed. Mayr also does not say what constitutes his "mountains of evidence" when clearly he should have—they are the fossils and inferences about fossils he calls "facts," which are accurately called "faith facts" and not facts about nature. Mayr believes that "evolution is a historical process"[62] and that "past stages cannot be observed directly, but must be inferred from the context."[63] Inferences are not facts, and they are not causal; they make nothing take place; they only exist in the mind. They certainly are not science. If they

21

cannot be observed directly, they must be imagined – through faith based inferences. Mayr's form of "proof" is testimonial and is worldview based. Mayr's claims immediately relegate natural selection to being no more than someone's beliefs in the form of non-causal inferences about fossils, fossil patterns, trees of life, and "empty of cause" fossil arrangements and classifications. They are "empty of cause" because no causal model is involved, physical nature is completely absent, but a worldview about nature is involved. Mayr knows that inferences are not causal or science or facts. The fossils are facts, however; and the fossils do not need to be refuted, for the fossils are there to be observed.

The fact of evolution is not in question. Evolution by natural selection is in question. The very fact that he claims "mountains of evidence" along with evolution as a causal process is no more than a testimonial, a voicing of his beliefs. Natural selection is not found alongside the fossils but in the mind, where it is not a fact of nature but a fact of faith that is found in Mayr's and others' "way of thinking" about the world. It is a way that is removed from science. Mayr discussed evolution, but like so many others, he did not claim "evolution by natural selection." Today, the term "evolution" typically means "evolution by natural selection," "evolution by God," "evolution by use and disuse," or some other causal model. Any model that is claimed to create new creatures is believed to be a cause of evolution, hence the reason for supporters of natural selection arguing against Christianity, Judaism, and other religions.

The National Academy of Sciences (NAS), BioLogos, and Evolution

The National Academy of Sciences (NAS) has written a thirty-nine-page, web-based booklet titled *Science and Creationism: A View from the National Academy of Sciences, Second Edition* (A View), which defines evolution as a process in the following way:

> The term Darwin most often used to refer to biological evolution was "descent with modification," which remains a good brief definition of the process today.[64]

The theistic evolution website, BioLogos, also defines evolution as "descent with modification." They write:

> The word *evolution* can be used in many ways, but in biology, it means *descent with modification.*[65]

The term "descent with modification" is additionally used as a synonym for "creation." For example, in 1872, Darwin had written about the veteran geologist M. J. d'Omalius d'Halloy, quoting him as stating "it is more probable that new species have been produced [created] by descent with modification than that they have been separately created." In his books, Darwin sees descent with modification performing acts of creation, causing evolution to take place. In the same way, he also sees natural selection perform those acts of creation. In effect, natural selection is equated to descent with modification thus to evolution. Why, one may wonder, are the terms being mixed with each other? Certainly, this is not scientific discourse or scientific definitions, but rather definitions that are used haphazardly. That practice is carried on to this day.

The NAS booklet, *A View*, defines evolution further:

> Evolution is a branching or splitting process in which populations split off from one another and gradually become different.[66]

Splitting is caused by modifications taking place and new creatures being formed from old ones, or rather "splitting" from the original parents. The NAS definition of evolution as a splitting process portrays it as the standard one used in discussing evolution by natural selection, called "anagenesis," a gradual smooth and continuous process whereby one species, over time, changes into a different species (a claim which conflicts with the fossil record which shows punctuated equilibrium or saltation). Another definition of a splitting process is where species become two or more distinct species by a process called "cladogenesis".

The NAS and BioLogos definitions of evolution follow a common, but incorrect, naming convention: evolution is defined as being the very process that caused it. That is, "evolution" is called by or used as "natural selection," which is also portrayed as "descent with modification." Darwin was guilty of confusing and mixing the terms involved with natural selection and he was criticized for defining evolution in the same way that is done by the NAS and BioLogos. He used "descent with modification" and natural selection as if they were the same. Samuel Butler, in his 1879 book, *Evolution Old and New*, writes:

> The difficulty, again, of understanding Mr. Darwin's mean-
> ing is enhanced by his repeatedly writing of "natural selec-
> tion," or the fact that the fittest survive in the struggle for
> existence, as though it were the same thing as "evolution" or
> the descent, through the accumulation of small modifications
> in many successive generations, of one species from another
> and different one.[67]

As Butler pointed out, Darwin, by the confusing use of his theory's key terms, "evolution" is defined as natural selection; evolution is also defined as descent with modifications, forming a circle of definitions: evolution equals natural selection equals descent with modification equals evolution. The confused use of these terms can be judged as cavalier or being indifferent to readers. The mixing of his terms may be held as invalidating those terms and, possibly, invalidating his theory of natural selection. Nature does not confuse its causal actions and the terms that represent nature cannot confuse the terms either, if it is to represent nature, yet that is what is taking place. Butler continues:

> "[O]n the theory of descent with modification" ... and in the
> next paragraph, "the theory of natural selection" is substi-
> tuted as though the two expressions were identical.

On the theory of the *natural selection* of successive, slight, but profitable, modifications, ... we find "*the theory of descent with modification*," and "*the principle of natural selection*," used as though they were convertible terms.[68]

Butler gives his view of the reason for this practice of confusing and mixing terms:

This is calculated to mislead. ... It leads the reader to forget that evolution by no means [evolution without a stated "cause"] stands or falls with evolution by means of natural selection, and makes him think that if he accepts evolution at all, he is bound to Mr. Darwin's view of it.[69]

As Butler was pointing out, every time a reader encounters evolution, they will think it refers to natural selection or descent with modification, which is a false association, but one which the mixing of terms leads them to believe. In reader's minds, evolution could then only be Darwin's way of thinking about it. Fisher, at the beginning of this chapter, pointed out, "Natural selection is not evolution." These definitions of evolution as "a process" incorrectly define it as natural selection, showing the NAS and BioLogos definitions of evolution to be incorrect: evolution is not the gradual creation of new body parts and new creatures; evolution does not work from a common ancestor. Evolution does not create anything at all for it is what is created. Darwin did not write, "evolution" was the main but not exclusive means of modification. He believed that natural selection causes "descent with modification" by variations being accumulated through natural selection, creating new body parts and new creatures, hence causing evolution to take place. The mixing of definitions has the effort of evolution being seen as natural selection, thereby, in the minds of readers, removing any chance of competing causes of evolution being considered. Darwin, it appears, has made an effort to have evolution, descent with modification, natural selection appear to be the same. The unscientific use of these terms is carried on to this day.

Darwin deliberately made the steady procession of chapters in his book appear as though it reflected the day-to-day progression of his research, giving an impression of a relatively uncomplicated progress from facts to ideas—but the real story is quite different.[70] As a historian of science, Janet Browne, Aramont professor of the history of science at Harvard University and editor of the *British Journal for the History of Science,* tells us:

> Natural selection was not self-evident in nature, nor was it the kind of theory in which one could say, "Look here and see."[71] Darwin had no crucial experiment that conclusively demonstrated evolution in action. He had no equations to establish his case. Everything in his book was to be in words—persuasion, revisualization, the balance of probabilities, the interactions between large numbers of organisms, the subtle consequences of minute chance and changes…he had to rely on drawing an analogy between what was known and what was not known, in Darwin's instance, by making a link between what took place in farmyards and what might be presumed to happen in the wild. He depended on probabilities. He relied on techniques in which the accumulation of factual examples progressively weakened a reader's resistance [this could be called proselytization].[72]

Browne gives an excellent account of the role of testimony used by Darwin. His *Origin of Species* was one long argument and not science. The same observations could be made for models used by the major religions: they are not self-evident in nature; one could not say, "Look here and see"[73]; and there are no experiments, no equations. Everything is in words (testimony)—persuasion, revisualization, analogies. Religious models of cause and effect have the same characteristics and operate in the same manner as Darwin's theories: using testimony.

Darwin's claims take the form of those of the major religions. In the *Origin of Species*, a reader commonly encounters "if", "I believe", "I think", "perhaps", "may be", "I suspect", "on this view", along with personifications, correlations, extrapolations, analogies, and metaphors – none of which are science and all of which are testimonial and causeless in nature. One major difference is that Darwin claimed his "theories" were science, ignoring the fact that there is no such thing as "testimonial science." But religion is always testimonial, as is the religion surrounding natural selection.

In place of independently repeatable observations, Darwin consistently argues to try to change the readers' way of thinking about nature. In other words, Darwin tries to replace the reader's belief system with his own beliefs. His attempt consists of "one long argument," which is a process of marketing and proselytization, not one of science. But it was the only process he had and also the only one that exists to this day: arguments consisting of inferences, extrapolations, correlations, guesses, and fiats. As Darwin writes in *Origin of Species*, he found case after case "inexplicable on the theory of independent acts of creation,"[74] as if the rejection of someone else's belief system had anything to do with the legitimacy of natural selection in nature or science. Darwin thought special creation, or independent creation, as he sometimes called it, left observations "inexplicable," which is not relevant, even if it were true. Darwin merely used his way of looking at the world (his worldview) to reject other worldviews. The cases Darwin cited as making "no sense" could make immense sense using another worldview. For many, creation and evolution using natural selection made no sense, a position held by his close friends Lyell and Huxley as well as Wallace (in some cases such as the creation of man's skin, hands, voice, and brain), the co-discoverer of natural selection who successfully tested natural selection's failure at creating some parts of people.

Summary

It seemed like science when John Tyndall, professor of natural philosophy at the Royal Institution and a member of Thomas Huxley's infamous X-Club, delivered the presidential address before the annual meeting of the British Association for the Advancement of Science, on the evening of August

19, 1874. Tyndall's topic was the relationship of science, past and present, to philosophical materialism.[75] One of the things taking place, however, was that evolution through natural selection was being testimonially attired as science. Tyndall's address was the occasion to state the aims and concerns of the premiere body of elite men of Victorian science. It was consequently one of the most prestigious places from which to opine on what men of science should be doing and thinking. In that momentous address, Tyndall was testifying that Darwin's evolution through natural selection was science, and he explained how it operated:

> He [Darwin] habitually retires from the more perfect and complex to the less perfect and simple, and carries you with him through stages of *perfecting*, adds increment to increment of infinitesimal change, and in this way gradually breaks down your reluctance to admit that the exquisite climax of the whole could be a result of Natural Selection.[76]

As is shown in Tyndall's above quote, natural selection, and therefore evolution through natural selection, operates entirely by testimony, not science. The early steps for natural selection and evolution being viewed as one concept, however inaccurately, were being taken. This was the beginning of evolution and natural selection being made to seem synonymous. Attributing observations to evolution by natural selection, as Tyndall was doing, is missionary-like testimony. It is testimony in its many forms: none of which is causal in nature or science.

Testimonies show how people think about creation but not how nature operates in the creation process. The testimonies show true believers' "argue and effect," not nature's "cause and effect." Tyndall, like Darwin, had as his goal the removal of the God of Genesis from people's "way of thinking" and believing about creation. It is a strange goal for a man said to be concerned with science. Darwin, who was being championed by Tyndall, as much admitted the removal of God from creation of new creatures as his goal.[77] Testimony about evolution through natural selection means creation without any God at all; it means preaching a religion, even if that religion is attired as science. Evolution was being fixed with the mask of natural selection despite evolution being an effect, not a cause, and despite its not having any relationship to natural selection, other than testimony. Through these early steps, today we often have the term "evolution" being erroneously used as if it were natural selection—wishful thinking of the faithful.

CHAPTER 2

MODELS OF EVOLUTION'S CAUSE

Darwin's book, *On the Origin of Species by Means of Natural Selection or the Preservation of Favoured Races in the Struggle for Life,* does not document the origin of a single species, or a single case of natural selection, or the preservation of one favored race in the struggle for life.[78]

—Jonathan Weiner, *The Beak of the Finch*

Introduction to Models of Evolution

Natural selection was not the first model claimed to cause evolution. All of Darwin's models, even natural selection, appeared at the tail end of the evolutionary parade of models and authors. There were many early models of evolution, such as preformation, an unfolding of a plan, fixity of the species, specific stability (punctuated equilibrium), special creation, pangenesis (Darwin's failed genetic theory), orthogenesis, need or appetency (as a strong desire, craving, instinct), intelligent design with God and without God, Wallace's God, gnostic God, life force, unknowable power, Darwin's God, and Darwin's wedging. The following list of five creation model examples presents different views of how new creatures were created, thereby causing evolution:

1. Preformation
2. Orthogenesis
3. Special creation

31

4. Wallace's God
5. Specific stability

A short description of each of these creation models shows how different views of creation caused evolution to taken place. They are presented to illustrate changes in thought over time.

1. Evolution by Preformation

In the past, evolution meant "jumping out of the box" and nothing more.[79] It could be called the "creation of all creatures at one time," with all life forms that would ever live being created instantly, in one act of *special creation*, then slowly being released at each birth. Jumping out of the box influenced the prevailing view of evolution, which was held to be "fixed," essentially unchanging. The view was one of *stasis*. Under the theory of "preformation," popular in the eighteenth century,[80] an individual develops by simple enlargement of a tiny, fully formed organism (called a homunculus) that exists in the germ cell. People existed, fully formed (as homunculi) from the start, within the ovaries. All creatures were preformed, and the concept of evolution was that creatures never changed into a new form with new organs and new systems of organs.

This eighteenth-century theory of how life produced copies of itself asserted that the tiny seeds or "germs" in a woman's body contain miniature versions of her offspring already fully formed, but too small to see, although some who looked through crude microscopes did believe that they saw tiny, preformed humans or "homunculi." Under this concept of creation, all the preformed humans do during pregnancy is grow larger, as they were already present in the mother's body when she was born. And her own "germ" was contained within her mother, "and so on back through the generations to the first woman, Eve."[81]

Explaining the preformation view of evolution further, Stephen Jay Gould tells us:

> [P]reformationists postulated a homunculus [small person] within the egg because they correctly understood that

a formless egg could not unerringly generate the same complex phenotype [body] again and again. In their world, form meant definite concrete structure—and they had no alternative but to postulate actual parts or organs of the next generation within the ovum [for nothing was known about genetics even through Darwin's time in the "1800s," up to the work of Gregor Mendel, who was a contemporary of Darwin's.[82]

Similarly, the reasoning that the fully formed offspring was inside the parent, like the butterfly that was contained in the chrysalis, the chrysalis in the caterpillar, the caterpillar in the ovum, the ovum in the butterfly, and so on[83] would *logically* lead back to one origin—the origin of all life made by the biblical Jewish and Christian creation model—God, who was the cause. Under the preformation approach to new creations, the ovum of two hundred billion human beings was once placed by the Creator in the womb of our first human mother Eve, in so light and diaphanous a form as to escape notice.[84]

No new forms of life were established under the concept of preformation, only the extraction of something that was already there:[85] one generation was set inside the other like a tiny image, similar to a set of Russian nesting dolls where one was opened and another was inside, which, when opened, contained still another identical but smaller doll inside. This backward procession went on and on to the original Adam and Eve and the creation of all creatures by God.[86] The model of creation was testimonially based. People merely believed in what had never actually been observed to take place, as with other creation models, including natural selection. Genetics was viewed differently in the eighteenth century. Even the most fundamental understanding was lacking. There was little advancement, even in Darwin's time, except for the work of Gregor Mendel, who created the first modern genetic model.

2. Evolution by Orthogenesis

Orthogenesis is the concept that life has an innate tendency to move in a *direction* that is driven by some internal "driving force." That internal driving force slowly transforms species into new creatures. Classic proponents

33

of orthogenesis have rejected the theory of natural selection, and opponents of orthogenesis accused its proponents of believing in "a mysterious inner force." At that time, paleontologists believed, practically as a united body, that variation has followed fixed lines through the ages—that there has been no such unrestricted and utterly free play (chance) of variational vagary as the Darwinian natural selection theory presupposes.[87] Despite that belief, one of the major challenges of orthogenesis was its vagueness. Paleontologists tended to accept "straight-line evolution" (orthogenesis) because they were aware of the "evidence" for long-continued trends. Straight line (rectilinear) creation was and is commonly observed, especially at the family and generic levels.[88]

The problem that existed (and still exists) with orthogenesis is that it could mean so many things.[89] Hence, it meant "everything" and "nothing." Orthogenesis is independent of nature, just like natural selection, miracles, special creation, descent with modification, and every part of every other creation model ever conceived. With orthogenesis, as with natural selection, creation takes place in the mind; one does not actually see new creatures gradually being created.

3. Evolution by Special Creation

Special creation may also be described as independent creation[90] or independent acts of creation. The Bible speaks to the way Christians and Jews accept special creation, and some of that detail is provided here for greater insight into its meaning. As with all testimonial creation models, special creation is accepted by faith as it does not have the characteristics of science models, where testimony is completely absent. Special creation has the same characteristics of acceptance or rejection as evolution by natural selection, which is by agreement or by disagreement. Faith is always involved with testimonial models that address origin, purpose, morals, and destiny, and they have none of nature's physical components. There are over one hundred references in the Bible for God's special creative activity.[91] Special creation is the first doctrine to be stated in Genesis:

In the beginning God created the heavens and the earth. (Genesis 1:1)

Special creation is also one of the last doctrines to be restated:

You are worthy, O Lord, to receive glory and honor and power; for You created all things, and by Your will they exist and were created. (Revelation 4:11)

The biblical account of the many separate acts of special creation teaches that creation was supernatural, which should raise the question of why Darwin argued against it when he had science on his side, or thought he did. Although the *effects* God created are physical and are found in nature, the causal acts of creation responsible for them remain unknown and rooted in faith. God states:

I am the Lord, who makes all things, who stretches out the heavens all alone. (Isaiah 44:24)

Creation was accomplished by the Word of the Lord, by the Word of the Lord the heavens were made, and all the host of them by the breath of His mouth. He gathers the waters of the sea together as a heap......For He spoke, and it was done; He commanded and it stood fast. (Psalm 33:6, 7, 9)

Special creation teaches that God created all that is seen from nothing. By His spoken word alone, the universe and all living creatures were created:

By faith we understand that the worlds were framed by the word of God, so that the things which are seen were not made of things which are visible. (Hebrews 11:3)

There was no imperfection in God's original creation. Imperfection eventually entered the universe as a result of mankind's sin, not God's design. Thus, the universe as it exists today is not the same as when God created it. Sin has brought into it abnormality and imperfection. Scripture speaks of God creating the parts of nature:

> For He looks to the ends of the earth, and sees under the whole heavens, to establish a weight for the wind, and mete out waters by measure, when He made a law for the rain, and a path for the thunderbolt. (Job 28:24–26)

The Bible contrasts the changelessness of God with an ever-changing creation:

> Of old You have laid the foundation of the earth, and the heavens are the work of Your hands. They will perish, but You will endure; yes all of them will grow old like a garment; like a cloak You will change them, and they will be changed. But You are the same, and Your years will have no end (Psalm 102:25–27).

The Bible says that God not only created the universe; He is also presently preserving it. Creation is dependent upon God.[92] The prophet Nehemiah writes:

> You alone are the Lord; You have made heaven, the heaven of heavens, with all their host, the earth and all things on it, the seas and all that is in them, and you preserve them all. The host of heaven worships You. (Nehemiah 9:6)

Jeremiah comments:

> He had made the earth by His power; He has established the world by His wisdom, and stretched out the heaven by His

understanding. When He utters His voice—there is a multitude of waters in the heavens; he causes the vapors of earth to ascend from the ends of the earth; He make lightning for the rain; He brings the wind out of His treasuries. (Jeremiah 51:15,16)

The apostle Paul says the creation gives the atheist no excuse as cause is linked to effect by apparent extrapolation to intelligence, values, knowledge, and purpose:

For since the creation of the world His invisible attributes are clearly seen, being understood by things that are made, even His eternal power and Godhead, so that they are without excuse. (Romans 1:20)

This claim is made by testimonial attribution, as Darwin did incessantly for natural selection. The testimonial claim is based on extrapolation, correlation, and inferences to an intelligent creator. Supernatural creation is an important biblical concept that is emphasized in both testaments. The testimony of Jesus refers to the creation account in Genesis.[93]

Have you not read that He who made them at the beginning made them male and female, and said, for this reason a man shall leave his father and mother and be joined to his wife, and the two shall become one flesh? (Matthew 19:4, 5)

In the above paragraph, Jesus quotes Genesis 1:27: "So God created man in His own image, in the image and likeness of God He created him; male and female He created them." He then quotes Genesis 2:24: "Therefore a man shall leave his father and his mother and shall become united and cleave to his wife, and they shall become one flesh."

Special creation has been exercised throughout the New and Old Testaments of the Bible. Darwin felt that the denial of special creation left only one creation model, natural selection. He shows this belief in *Origin of Species* by arguing as if creation is either by special creation or natural selection. According to him, special creation is an exercise of faith, whereas natural selection is science. However, it is not one or the other (natural selection or special creation) as any causal model stands or falls on its own merits, especially if it is a science model. Natural selection *falls* because it has nothing of nature in it, just like special creation has nothing of nature in it. Natural selection has testimonial links between cause and effect, just like special creation has testimonial links between cause and effect. Natural selection operates by testimonial attributions being made to it, just like special creation has testimonial attributions made to it. Natural selection may be accepted or rejected, just like special creation may be accepted or rejected. Natural selection cannot be used to do anything in nature, just like special creation cannot be used to do anything in nature, such as develop technology. The two creation models, natural selection and special creation, are the same types of models: testimonial. There is no science to either one. Both exist as exercises of faith.

4. Evolution by Wallace's God

Wallace's God was different from Darwin's God, the deist God, and the Judeo-Christian God. According to Wallace, his God only created certain parts of humans. The difficulty with Wallace's God is that His identifying characteristics are not clear. As shown in chapter 11 in this book, "Testing Miracles and Natural Selection," Wallace's tests prove that God created humans' higher qualities, and natural selection did not. He did not think the name or type of God to be important. He did not seem to be overly concerned "whether we call it God or spirit."[94] He did care that his God played an important role in human evolution.[95] Wallace writes the following about God:

> We thus find that the Darwinian Theory, even when carried out to its extreme logical conclusion, not only does not oppose, but lends a decided support to, a belief in the spiritual nature of man. It shows us how man's body may have been developed from that of a lower animal form under the law of natural

selection; but it also teaches us that we possess intellectual and moral faculties which could not have been so developed, but must have had another origin; and for this origin we can only find an adequate cause in the unseen universe of Spirit.[96]

Wallace's God is not necessarily viewed as the Creator, but more like an eternal animating force behind the universe, with the universe as nothing more than the manifest part of God. Wallace considered his concept of God to be scientific but could not change the fact that the link from cause to effect was testimony. He merely exchanged one belief system about God for another, which is still outside the realm of science.

5. Evolution by Specific Stability

Although historians of science generally recognize the 1972 paper on punctuated equilibrium by Eldredge and Gould as the foundational document of the new paleobiological research program, in 1971, Eldredge published a paper in the journal *Evolution* suggesting that gradual evolution was seldom seen in the fossil record. He argued that Ernst Mayr's standard mechanism of allopatric speciation might suggest a possible resolution.[97] However, punctuated equilibrium, under a different name, was discussed over one hundred years before Eldredge and Gould published their 1972 paper.[98] Using the term "specific stability," Mivart writes of rapid creations and stability (stasis) of creatures:

> Thus, then, it seems that a certain normal specific stability in species [stasis], accompanied by occasional sudden and considerable modifications [punctuation], might be expected.[99]

> Arguments may yet be advanced in favor of the view that new species have from time to time manifested themselves with *suddenness*, and by modifications appearing at once (as great in degree as are those which separate Hipparion from Equus), the species remaining stable in the intervals of such modifications.[100]

Stability (stasis) is defined thus:

> [B]y *stable* being meant that their variations only extend for a certain degree in various directions, like oscillations in a stable equilibrium.[101]

These variations are the normal distributions of body parts about an average or mean showing that nothing is changing. Any group of creatures, including humans, show these normally distributed variations. The finch beaks on the Galapagos illustrate this normally distributed oscillation about a mean, with nothing new being created. In an 1871 review of Mivart's book, *Genesis of Species,* reviewer Chauncey Wright also defines specific stability with more depth:

> It [the creature] passes, according to the hypothesis, from one form to another of specific "manifestation," abruptly and discontinuously in conformity to the emergencies of its outward life; but in any condition to which it is tolerably adapted it retains a stable form, subject to variation only within determinate limits, like oscillations in a stable equilibrium.[102]

Mivart shows that non-mutational variations oscillate about fixed body parts and a body plan, like variations in humans' height and sizes of hats, shoes, or gloves, but never go beyond a certain amount. With this concept, body parts from parent to offspring vary, but never change into something else; further, the variations never accumulate. In other words, a mouse does not become an elephant, or a bear does not become a whale. Mivart continues with stability, explaining its meaning:

> [S]pecies are stable at least in the intervals of their comparatively sudden successive manifestations.[103]

Being stable means oscillating in equilibrium about a common body plan and architecture; it means stasis, or remaining the same. Mivart uses different names to describe punctuated equilibrium. His phrase "sudden successive manifestations"[104] means "punctuated"; and his phrase "variations…like oscillations in a stable equilibrium"[105] means "the dynamic equilibrium of variations about a common body plan." In other words, offspring have different dimensions, weights, sizes, volumes, and so on, but within limits and never to exceed those limits. The oscillations take the form of a normal distribution. Mivart and Galton show, even in 1871 and before, that fossils were acknowledged to appear suddenly and remain stable afterward, as shown again in 1972 with Eldred and Gould's punctuated equilibrium. Francis Galton, Darwin's half cousin, writes of his description of punctuated equilibrium (then called specific stability) using an analogy of a "many-facetted spheroid tumbling over from one facet, or stable equilibrium, to another." In his 1869 book, *Hereditary Genius*, Galton writes:

> I will now explain what I presume ought to be understood, when we speak of the *stability* of types, and what is the nature of the changes through which one type yields to another. Stability is a word taken from the language of mechanics; it is felt to be an apt word; let us see what the conception of types would be, when applied to mechanical conditions.[106]

> The metaphoric mechanical conception of stability would be that of a rough stone, having, in consequence of its roughness, a vast number of natural facets, on any one of which it might rest in "stable" equilibrium. That is to say, when pushed it would somewhat yield, when pushed much harder it would again yield, but to a lesser degree; in either case, upon the pressure being withdrawn, it would fall back into its first position.[107]

But, if by a powerful effort the stone is compelled to overpass the limits of the facet on which it has hitherto found rest, it will tumble over into a new position of stability, whence just the same proceedings must be gone through as before, before it can be dislodged and rolled another step onward. The various positions of stable equilibrium may be looked upon as so many typical attitudes of the stone, the type being more durable as the limits of its stability are wider. We also see clearly that there is no violation of the law of continuity in the movements of the stone, though it can only repose in certain widely separated positions.[108]

Galton aptly illustrated punctuated equilibrium using the metaphor of a rough stone with many facets purely as an aide in trying to convey what he thought was taking place. This too was punctuated equilibrium. While Galton's metaphor is not science, his view showed that new bodies cannot be created gradually, without every part of the system immediately providing all the necessary functions of life, such as sustainment (bringing in food and removing wastes); maintenance (repairing and replacing parts); operations (by the nerves connected to all the parts of the body); reproduction; control-feedback systems; and autonomic and somatic (conscious) control systems to which all the operations would be assigned by the creation model. This would include immensely complex systems such as identification systems, transportation systems, communications systems, body temperature control systems, chemical production systems, reproduction systems, and rules of operation. Every creature needs these essential systems to come to life and remain alive. The creation model that meets the requirements of sudden creation and complete creation of these essential body systems would be a candidate for the creation model of creatures' bodies.

Summary

Each proposed model of creation in all of evolutionary biology exists independently of nature. That is, every creation model ever proposed lacks the identified physical parts of nature that cause creation to take place, such as biological materials, shapes, dimensions, locations, and their relationships that operate for life's sustainment. Without these components of nature, the creation model is a mere reference point, a platform for testifiers such as Darwin. Harvard's Ernst Mayr tells us that "evolution is a historical process and that past stages cannot be observed directly and must be inferred."[109] Apparently Mayr overlooks the fact that inferences have useful characteristics, but being scientific is not one of them. He and others in evolutionary biology do not see the fact that inferences are not causal and do not exist in nature, disqualifying them from being representative of nature's physical processes or having the characteristics of a model of science.

Natural selection was proposed as a causal model of evolution, a creation model that is believed to be responsible for creation of creature's body parts. Inferences do not create anything but thoughts in a person's mind. If inferences are to be science, then all religions become science. To be science, a model must show how effects are caused, meaning that a body's identified parts and systems must be shown in the creation model performing the creation of the countless creatures' body parts. What is shown with all the models of evolutionary biology, including natural selection, is that they do not physically exist in nature and require an abundance of testimony to supply the endless attributions of imagined creations taking place.

Natural selection does not contain any parts of giraffe necks, or whales created from land animals, or peppered moths (which never have genetic changes), or mimicry; so it is easy to see why Darwin, Wallace, Dobzhansky, Mayr, and the NAS must use inference (which never actually causes anything to take place)— it is the biggest tool in their toolbox. It should be remembered that testimony, of which inference is a part, has no merits in nature, whether it be from Darwin, Wallace, Dobzhansky, or the NAS. Testimony adds faith, not fact, to the public debate. A large part of the testimony consists of non-causal imaginary models such as arms races, adaptive landscapes, selection pressures, gradual creation, and game theory. Testimonies such as these are likely given with the hopes of showing others "a way of thinking" about creation that may be different than their own, perhaps influencing them to adopt it as their faith system.

43

CHAPTER 3

DARWIN AND PRIORITY

In England, if not everywhere, most botanists and zoologists were a muddled lot. Not even the possession of University Chairs gave many the assurance to do clean science; some actually wasted their efforts on useless polemics about the origin of life or how we know that a scientific fact is really correct.[110]

—James D. Watson, Nobel Prize winner

Introduction

A parade of people had written about evolution before Darwin. Their work paved the way for Darwin's publication of the 1859 *Origin of Species*,[111] wherein he used those ideas. Some historians believe that Darwin borrowed *all* of the major contributions he is credited with, including natural selection, from other naturalists. Some use the word "plagiarized." Many, if not most, of Darwin's major ideas exist in earlier works, including those by his grandfather, Erasmus Darwin. Charles Darwin rarely (if ever) gave due credit to the many persons from whom he liberally "borrowed." Due to the pressure of being accused of using other people's ideas, Darwin eventually recognized the contributions of others in the 1861 third edition of *Origin of Species*, wherein he made reference to many of these men and gave his view of their contributions. He carried that acknowledgment in the "Historical Sketch" at the beginning of the third, fourth, fifth, and sixth editions of his book.

45

In Samuel Butler's 1882 book, *Evolution Old and New,* Butler reviewed the contributions of others that preceded Darwin, such as Buffon and Erasmus Darwin. Butler charged Darwin with plagiarizing. Darwin had not published any work on evolution or natural selection prior to the 1859 *Origin of Species*; and in that first edition, he did not even mention the fact that, just prior to his publishing his 1859 book, Wallace had sent Darwin a complete work on divergence and natural selection known as the "Ternate Essay," so named because it was written on the island of Ternate; the essay was titled, "On the Tendency of Varieties to Depart Indefinitely from the Original Type." This essay was written in February 1858 and sent directly to Darwin with the expressed wish that it should be forwarded to Lyell. Natural selection and evolution were spelled out in this paper.

In the *Origin of Species*, Darwin continued his practice of withholding credit by not mentioning the Ternate Essay which Wallace had sent to him directly. That paper caused an explosion under Darwin that moved him to hurriedly publish his book, *Origin of Species*, forgetting to mention all those whose work he had read or "borrowed." Darwin only made a meager mention that Wallace had researched and published an 1855 "Sarawak law" on evolution whose full title is, *"On the Law Which Has Regulated the Introduction of New Species."* Wallace's publication of that law preceded Darwin's publication of *Origin of Species*. On page 355 of *Origin of Species,* Darwin buried acknowledgment of Wallace's seminal evolutionary work by referencing a line from Wallace's Sarawak work, "[E] very species has come into existence coincident both in space and time with a preexisting closely allied species,"[112] which conveys Wallace's first formal statement of his understanding--a pre-natural selection understanding--of the process of biological evolution, *published in Volume 16 (2nd Series) of the Annals and Magazine of Natural History in September 1855.*

In his 1901 book, *Lamarck, the Founder of Evolution His Life and Work,* Alpheus S. Packard writes about Charles Darwin's use of his grandfather's writings about evolution:

The grandson of Erasmus Darwin had little appreciation of
the views of him of whom, through atavic heredity, he was the

intellectual and scientific child. "It is curious," he says in the 'Historical Sketch' of the *Origin of Species*—"how largely my grandfather, Dr. Erasmus Darwin, anticipated the views and erroneous grounds of opinion of Lamarck in his *Zoonomia*, published in 1794."[113]

Darwin had read his grandfather's work but gave no reference to it. Packard continues:

> It seems a little strange that Charles Darwin did not devote a few lines to stating just what his ancestor's [grandfather's] views were, for certain of them, as we shall see, are anticipations of his own.[114]

Simple courtesy, correctness of priority, and family pride should have forced Charles Darwin to publicly acknowledge his grandfather's ideas, but he failed to do so. In 1858, one year before the publication of *Origin of Species*, Darwin had already become the most famous scientist of his day. For twenty years, he had struggled unsuccessfully to formulate a complete explanation for evolution, but had not published so much as a line on the subject.[115]

In his 1980 book, *A Delicate Arrangement*, Arnold C. Brackman reveals how Darwin and two friends conspired on Darwin's behalf to secure priority and credit for the theory of natural selection.[116] Darwin had been corresponding with Alfred Russel Wallace about evolution and its cause. After receiving a letter from Alfred Russel Wallace (then an unknown, penniless, self-educated naturalist and adventurer), the wealthy and well-connected Darwin saw before him an explanation, a theory that spelled out everything in detail: Wallace's 1858 "Ternate Paper," titled "On the Tendency of Varieties to Depart Indefinitely from the Original Type." Brackman presents the case for Darwin taking credit for Wallace's work and taking sole priority from Wallace. Wallace had written about geographic and geologic distribution of both living creatures and fossil species in what would become known as biogeography. His conclusion has since come to be known as the Sarawak law,[117] which claims:

> Every species has come into existence coincident both in space and time with a closely allied species.[118]

In the 1859 *Origin of Species*, Darwin did not acknowledge Wallace as having published the "Sarawak law" in 1855, which Darwin had read four years before he published *Origin of Species*. In his posthumously published 1995 book, *Darwinian Fairytales*, Australian philosopher of science David Stove (1927–1994) writes of Darwin's lifelong habit of not acknowledging the debts his work owed to other people (at least not until he was obliged to do so); he had a still worse habit of not even noting the people whose ideas he "borrowed."[119] During Darwin's lifetime, Stove's accurate portrayal of Darwin was also presented by Thomas Sims, Wallace's brother-in-law, who felt that Darwin slighted Wallace on the theory of natural selection by barely acknowledging Wallace (discussed in the summary of this chapter). In her 2002 book, *The Power of Place*, the historian Janet Browne writes that Darwin depended on the postal system of his day for the collection of information (the modern Internet of his day): "The flow of information that was initiated was almost always one-way."[120] It flowed from others to Darwin. Browne tells us that Darwin regarded his correspondence primarily as a supply system, designed to answer his own wants. Among his closest friends, however, Darwin proved unwilling to appear quite so exploitative.[121]

During the five-year voyage of the *Beagle*, Charles Darwin read and used Lamarck's *Histoire Naturelle des Animaux sans Vertèbres* as well as the second volume of Lyell's *Principles of Geology*. Reading these works on evolution by various causes presented Darwin with much material for his own writing during the voyage. Darwin's ideas about island speciation, evolution, use and disuse, survival of the fittest, common ancestor, descent with modification, and others came into being as much because of what he read on the *Beagle* as any finding he made because of his voyage. Darwin was not a revolutionary, original thinker, despite mistaken claims that he revolutionized the concept of evolution by introducing natural selection. Darwin was, quite apparently, a slow organizer of other people's materials. He spent many years forming a well-documented testimonial case for a creature's offspring gradually changing into new creatures, which he then presented as "evolution by natural selection." The idea of selection was not even put forth by him, as others had already cited animal breeders selecting creatures for breeding. He was also not the first to write about natural selection. It is clear that, by accomplishments alone, Darwin did not represent the greatest biologist before, during, or after his time. As discussed below, Pasteur and Mendel deserve that recognition for their brilliant work in the biology of vaccinations, pasteurization, spontaneous generation, germ theory, immunization, and genetics. There were, in fact, many people whose accomplishments in biology and science proved more significant and useful than natural selection, which, as shown in this work, *is inoperative in nature and contains no biology.*

This chapter will show that Darwin was not the first to accomplish anything of note regarding evolution. Darwin did not bring one new concept to the parade of evolutionary ideas. He was not the first to: popularize evolution, discover natural selection, write about descent with modification, use chance creation, use variations or direction, invent survival of the fittest, use competition, use war of nature, cite gradualism, discover sexual selection, use common ancestor, use island biogeography, use the analogy of animal breeders, invent the tree of life, discover morphology, write about the ape to man transition, or use the term evolution. A brief list of a few of these items to which Darwin has no priority follows.

1. popularize evolution,
2. publish, discover, or use natural selection,
3. use "survival of the fittest," or
4. use the animal breeders' analogy.

Notable among the items in this list is that a total of four men published their work about natural selection before Darwin (James Hutton, William Wells, Patrick Matthews, and Edward Blyth). Added to these four is a fifth, Alfred R. Wallace. If Wallace's "Ternate Paper" that he sent to Darwin is counted as public disclosure or publication (for Darwin, it was), then Wallace was the fifth person to publish or publicly release information about natural selection. The descriptions of the above four numbered items follow with each description identified by matching the numbers above:

1. Darwin Was Not the First to "Popularize Evolution"

The successful sales of many editions of Robert Chambers' anonymously published *Vestiges of the Natural History of Creation*, beginning in 1844 (fifteen years before Darwin's *Origin of Species*) paved the way for Darwin's sales. Darwin could not have helped but know that, indeed, many among the general public would read his book,[122] if for no other reason than because the subject matter mimicked that of Chambers. Chambers attained the unprecedented popularity that paved the way for Darwin. Chambers' *Vestiges* helped accustom more people to the claimed mechanisms of naturalistic evolution. It was widely read, not only by members of high society but also by the lower and middle classes, thanks to the rise of less expensive publishing methods. *Vestiges* continued to sell in large quantities for the rest of the nineteenth century.[123]

The first edition of *Vestiges* sold 1,750 copies and sold out in a few days. Among those fortunate enough to have ordered copies promptly, Lord Alfred Tennyson commented to his bookseller that it "seems to contain many speculations with which I have been familiar for years, and on which I have written more than one poem."[124] Having read the book, Tennyson concluded, "There was nothing degrading in the theory."[125] Benjamin Disraeli (1804–1881) was a British conservative politician, writer, and aristocrat who twice served as Prime Minister of England.[126] He told his sister that the book was "convulsing the world,"[127] and his wife told her that "Dizzy says it does and will cause the greatest sensation and confusion."[128]

Vestiges presented a progressive "law" with humanity as its goal (and thus continuity), which treated the human race as the last step in the ascent of animal life. It argued that mental and moral faculties were not unique to humans but resulted from expansion of brain size during this ascent.[129] Darwin acknowledged that *Vestiges* "prepared the ground for the reception of analogous views" such as his. In the 1861 third edition of the *Origin of Species*, Darwin finally writes about *Vestiges*:

The "Vestiges of Creation" appeared in 1844. In the tenth and much improved edition (1853) the anonymous author says... The work, from its powerful and brilliant style, though displaying in the earlier editions little accurate knowledge and a great want of scientific caution, immediately had a very wide circulation. In my opinion it has done excellent service in calling in this country attention to the subject, in removing prejudice, and in thus preparing the ground for the reception of analogous views.[130]

Using his own definition of "evolution," Hugo de Vries names four men who published notable works about evolution. De Vries, the coiner of the term "mutation" and one of the men who rediscovered Gregor Mendel's work, notes those who wrote about evolution before Darwin when he writes the following:

Evolution, meaning the origin of new species by variation from ancestor species, as an explanation for the state of the living world, had been proclaimed before Darwin by several biologists/thinkers, including the poet [1] Johann Wolfgang Goethe, in 1795, [2] Jean-Baptiste de Lamarck in 1809, [3] Darwin's grandfather, the ebullient physician-naturalist-poet philosopher Erasmus Darwin, and in Darwin's time anonymously by [4] Robert Chambers in 1844.[131]

Darwin was not the first to publish views about evolution, whether by naturalistic or any other causes. Note that the "origin of new species by variation from ancestor species" translates to "common ancestor," "descent with modification through some cause," and "naturalistic evolution." The terms used by Darwin may not have been used by earlier writers about evolution's cause, but the concepts had been presented, such as evolution, descent through modification, survival of the fittest, and sexual selection.

2. Darwin Was Not the First To: Publish, Discover, or Use Natural Selection.

Darwin is often perceived as making the groundbreaking discovery of natural selection, which is an incorrect view. He was not the first. This section presents the men who published works referencing natural selection prior to Darwin. In 1794, James Hutton, known as the father of modern geology, referenced natural selection but did not use that term. In chapter 3, section 13, book 2 of *The Principles of Knowledge*, his writing references both survival of the fittest and natural selection. For example, Hutton writes:

> This wisdom of nature, in the seminal *variation* of organised bodies, is now the object of our contemplation, with a view to see that the acknowledged variation, however small a thing in general it may appear, is truly calculated for the preservation of things, in all that perfection with which they had been, in the bounty of nature, first designed. Now, this will be evident, when we consider, that if an organised body is not in the situation and circumstances *best adapted* to its sustenance and propagation, then, in conceiving an indefinite variety among the individuals of that species, we must be assured, that, on the one hand, those which depart most from the *best adapted* constitution, will be most liable to perish [unfit], while, on the other hand, those organised bodies, which most approach to the best constitution for the present circumstances, will be *best adapted* [fittest] to continue, in preserving themselves and multiplying the individuals of their race. [Italics added][132]

That is survival of the fittest and natural selection, complete with variations. After Hutton's work, and forty-five years before Wallace and Darwin had their works read before the Linnean Society, William Wells (1757–1817) delivered his paper before the Royal Society of London. In it he observed that "amongst men, as well as among other animals, varieties of a greater or less magnitude are constantly occurring." Wells attributed this to natural selection and the survival of the fittest.[133] The scientific topics that Wells addressed ranged over many areas of research. Wells wrote about evolution by natural selection and survival of the fittest, well ahead of Spencer and Darwin. In an article by C. R. P. George, Renal Physician, Concord Hospital, NSW Australia, published in "Nephrol Dial Transplant (1996)", European Renal Association-European Dialysis and Transplant Association, George says of the scientific topics Wells addressed:

> The scientific topics that he addressed ranged from investigations upon why humans have single vision with two eyes (1792); the mechanism of electrical stimulation of muscles (1795); the reason for the red coloration of the blood (1797); the mechanism of pupillary reactions (1811); the implications of coloration of the skin, with proposal of a theory of *evolution as natural selection by survival of the fittest* (1813—in this long preceding Charles Darwin); to the mechanism of the formation of dew (1814). Most of these publications appeared in Philosophical Transactions of The Royal Society. [Italics added][134]

Darwin acknowledged others that had priority before him (including Wells), in the "Historical Section," at the beginning of the 1866 fourth edition of *Origin of Species*. Darwin writes in the 1866 *Origin of Species*:

> In 1813, Dr. W. C. Wells read before the Royal Society "An Account of a White Female, part of whose skin resembles that of a Negro"; but his paper was not published until his famous "Two Essays upon Dew and Single Vision" appeared in 1818. In this paper he distinctly recognizes the principle of natural selection, and this is the first recognition which has been indicated; but he applies it only to the races of man, and to certain characters alone...[H]e observes, firstly, that all animals tend to vary in some degree, and, secondly, that

agriculturists improve their domesticated animals by selection; and then, he adds, but what is done in this latter case "by art, seems to be done with equal efficacy, though more slowly, by nature, in the formation of varieties of mankind, fitted for the country which they inhabit. Of the accidental varieties of man, which would occur among the first few and scattered inhabitants of the middle regions of Africa, someone would be better fitted than the others to bear the diseases of the country. This race would consequently multiply, while the others would decrease; not only from their inability to sustain the attacks of disease, but from their incapacity of contending with their more vigorous neighbours. The colour of this vigorous race I take for granted, from what has been already said, would be dark. But the same disposition to form varieties still existing, a darker and a darker race would in the course of time occur: and as the darkest would be the best fitted for the climate, this would at length become the most prevalent, if not the only race, in the particular country in which it had originated." He then extends these same views to the white inhabitants of colder climates. I am indebted to Mr. Rowley, of the United States, for having called my attention, through Mr. Brace, to the above passage in Dr. Wells' work.[135]

Darwin writes more voluminously, but not differently, on key points. In the above, Wells describes selection, fitness, and gradual evolution. Darwin's account of natural selection reads with the same key points as those of Wells. It is Paul N. Pearson, in an article in *Nature*, who gives precedence to Wells for first discovering natural selection and "honorable mention" to Hutton.

Credit for the first appreciation of natural selection could therefore go to Wells rather than to Edward Blyth or Patrick Matthew. The triumph is limited to the extent of being applied only to skin colour, and not, as Darwin and Wallace did, to the whole range of life. A form of the idea had already been set out by an earlier Edinburgh author, James Hutton, but in that case the effect was limited to improvement of varieties rather than the formation of new species.[136]

Charles Hodge (1797–1878) was the principal of Princeton Theological Seminary between 1851 and 1878. In his 1874 book, *What Is Darwinism*, Hodge shows that Darwin does not have priority for natural selection.

> Mr. Darwin not only says that he had been anticipated in teaching the doctrine of evolution by Lamarck and the author of the "Vestiges of Creation"; but that the theory of natural selection, as the means of accounting for evolution, was not original with him.[137] He [Darwin] tells us that as early as 1813 W. C. Wells 'distinctly recognizes the principle of natural selection"; and that Mr. Patrick Matthew, in 1831 "gives precisely the same view of the origin of species as that propounded by Mr. Wallace and myself."[138]

Darwin grudgingly gives credit to Wells and Matthew's for evolution and natural selection. In his 1882 book, *Evolution, Old and New*, Samuel Butler writes that Matthew's work was to be "one of the most perfect yet succinct expositions of the theory of evolution that I have ever seen."[139] Butler had read Charles Darwin's account of natural selection as well as Erasmus Darwin's account of evolution, as written in Erasmus Darwin's 1796 first-edition book, *Zoonomia*. Butler praises Matthew's 1832 presentation of evolution by natural selection.

> This must be connected with Mr. Matthew's work on "Naval Timber and Arboriculture," which appeared in 1831. The remarks which it contains in reference to evolution are confined to an appendix, but when brought together, as by

Mr. Matthew himself, in the "Gardeners' Chronicle" for April 7, 1860, they form one of the most perfect yet succinct expositions of the theory of evolution that I have ever seen. I shall therefore give them in full. This [Matthew's] book was well received, and was reviewed in the "Quarterly Review," but seems to have been valued rather for its views on naval timber than on evolution. Mr. Matthew's merit lies in a just appreciation of the importance of each one of the principal ideas which must be present in combination before we can have a correct conception of evolution, and of their bearings upon one another. In his scheme of evolution I find each part kept in due subordination to the others, so that the whole theory becomes more coherent and better articulated than I have elsewhere found it [such as in Charles Darwin's 1859 *Origin of Species*].[140]

Samuel Wilberforce, bishop of Oxford, writes in his 1860 review of Darwin's first edition of *Origin of Species* that many of the ideas Darwin presented involving evolution were written by his grandfather, Erasmus Darwin. Charles Darwin, who had read his grandfather's book, seems to have forgotten to give credit to his grandfather. Wilberforce writes:

For if we go back two generations we find the ingenious grandsire [grandfather, Erasmus Darwin] of the author of the 'Origin of Species' speculating on the same subject, and almost in the same manner with his more daring descendant.[141]

Beneath the surface of public consciousness today, and almost never mentioned by those who should or do know, lies the incorrect attribution of priority to Darwin for many of the concepts involved with evolution. We may only wonder why. Darwin has no priority with evolution, natural selection, or survival of the fittest. Priority goes to the first person to publish a concept. Darwin includes Matthew among the thirty-four authors in his Historical

Sketch at the beginning of the 1861 third edition of *Origin of Species*,[142] but he omitted these authors in the first two editions.

The Darwin exhibition, created by the American Museum of Natural History, was the centerpiece of the bicentennial of Darwin's birth. It opened in November 2005 and circulated to a number of museums before terminating at the London Natural History Museum in February 2009.[143] Hutton, Matthews, Wells, and Blyth were not mentioned in the American Museum of Natural History's Darwin Centennial,[144] despite having written about natural selection prior to Darwin. In that exhibition, Darwin served as the centerpiece, but accuracy of facts about evolution, priority, and others' contributions to biology (Pasteur) and genetics (Mendel) did not factor into the exhibition, despite some of those men being contemporaries of Darwin. The importance of Darwin's work pales, if not evaporates, when compared to those men and others, both prior to him and of his own time.

Even though Matthew is mentioned by Darwin in the third edition of the *Origin of Species* as deserving credit for natural selection, the Darwin Exhibition makes no mention of Matthews, which, if it is not intentional, is an oversight that is not understandable, given the education and positions of those that are or should be aware of the history involved. Any professional organization, such as the American Museum of Natural History—especially when its study of nature is devoted to natural selection as its centerpiece—should have given Hutton, Wells, or Matthew credit of priority, or, if not convinced of their priority, should have at least mentioned Matthew's and other predecessors work and contributions. It appears to be a glaring omission for an organization of such presence in the worldwide evolution community. On page xiii of the *Origin of Species'* 1861 third edition, Darwin recognizes Matthew's priority in the section titled "An Historical Sketch of the Recent Progress of Opinion on the Origin of Species." Perhaps the people the American Museum of Natural History did not read what Darwin had written:

> In 1831 Mr. Patrick Matthew published his ideas on "Naval Timber and Arboriculture," in which he gives precisely the same view on the origin of species as that (presently to be

alluded to) propounded by Mr. Wallace and myself in the "Linnaean Journal," and as that enlarged on in the present volume. Unfortunately the view was given by Mr. Matthew very briefly in scattered passages in an Appendix to a work on a different subject, so that it remained unnoticed until Mr. Matthew himself drew attention to it in the excellent history of opinion on this subject.[145]

Darwin also gives credit to Wells for his discovery of natural selection in the "Historical Sketch" at the beginning of the fourth, fifth, and sixth editions of *Origin of Species*, where he writes:

In 1813 Dr. W. C. Wells read before the Royal Society "An Account of a White Female, part of whose Skin resembles that of a Negro"; but his paper was not published until his famous "Two Essays upon Dew and Single Vision" appeared in 1818. In this paper he distinctly recognizes the principle of natural selection, and this is the first recognition which has been indicated; but he applies it only to the races of man, and to certain characters alone.[146]

Darwinism could actually be called "Wellsism." In another instance, Patrick Matthew discovered natural selection before Darwin. As with Darwin and Wallace, Matthew even claimed that Thomas Robert Malthus (1766–1834) was the key incentive to the discovery of natural selection. Matthew had read Malthus's book, *An Essay on the Principle of Population.*[147] Malthus was a British demographer and political economist, best known for his highly influential views on population growth. The priority of natural selection's discovery goes to Hutton (first), Wells (second), then Matthew (third). As Darwin recognized Matthew, Darwinism may rightly be called "Matthewism."

In 1836, Edward Blyth, the fourth person to publish about natural selection, published three closely argued natural history essays in which he introduced the concept of natural selection. Edward Blyth wrote three articles on variation, discussing the effects of artificial selection (as Darwin did) and describing the process in nature (later called natural selection) as restoring organisms in the wild to their archetype (rather than forming new species[148]). However, he never actually used the term "natural selection." These articles were published in the *Magazine of Natural History* between 1835 and 1837.[149] Blyth, long a member of Darwin's circle of sources, endorsed evolution (by natural selection) and was acknowledged by Darwin after publication of *Origin of Species*. The American Museum of Natural History's exhibition toured different countries and made no mention of Edward Blyth. The second passage is taken from Darwin's *Origin of Species* (1859) and the first passage is from the first of Blyth's papers, "The Varieties of Animals," published twenty-four years earlier in 1835 in the *Magazine of Natural History*.[150]

> "'True Varieties'...what are, in fact, a kind of deformities, or monstrous births [saltations]...would very rarely, if ever, be perpetuated in a state of nature.
>
> —Blyth

> It may be doubted whether sudden and considerable deviations of structure [saltations] such as we see in our domestic productions...are ever permanently propagated in a state of nature.
>
> —Darwin

The two statements contain the same basic idea: saltations may not survive in nature, but room is left for the possibility that such sudden creations may survive and lead to new creatures. Darwin corresponded with Blyth, who was in India during the correspondence. Making the case for plagiarism, a detailed paper about Blyth and Darwin was written by Andrew J. Bradbury, with the title, "Edward Blyth—Did Darwin Plagiarize from Blyth?"[151]

Wallace's "Sarawak Law" showed his thoughts on evolution up to that point. This paper, written at Sarawak in Borneo in February of 1855 and published in volume 16 (second series) of the *Annals and Magazine of Natural History* in September 1855, conveys Wallace's first formal statement of his understanding, a prenatural selection understanding of the process of biological evolution.[152] For years, he had pursued evolution, chasing after its cause, forming a concept of how creatures were created. But Wallace's crowning discovery and writing about natural selection came from a rented home on Ternate,[153] an island in the Dutch East Indies halfway between Celebes and New Guinea.[154] Wallace took several days[155] to write a paper on evolution by natural selection in February 1858.[156] He then sent the "Ternate Paper" to Darwin, asking him to read it and to pass it to Lyell, who was close friends with Darwin. Upon receipt of the "Ternate Paper," Darwin felt mortified, as he saw his priority gone. Lyell recommended that Darwin publish a short response to Wallace's "Ternate Paper"[157] and have it publicly read to prevent Wallace from being recognized as having priority. In this way, both Wallace and Darwin would publicly become co-discoverers. But when Darwin hurriedly published his 1859 first-edition *Origin of Species*, he mentioned neither the "co-discovery" nor Wallace's Ternate letter to Darwin.

Lyell and Darwin chose to go before the Linnean Society and read Darwin's and Wallace's papers without Wallace's knowledge[158]. This was not what Wallace had requested from Darwin or intended from Lyell. Had Wallace sent it directly to a publisher, his priority would have been cemented in history. However, Darwin wrote a hurried response to Wallace's paper,[159] and he "dispatched some 1844 manuscript material about evolution" to Hooker on 29 June [1858]."[160] Darwin's response to Hooker was a very hasty culling of paperwork. The content of the package to Hooker was not from a manuscript that was to be sent to a publisher; rather, it was an odd mixed bundle for the purpose of priority, a very hasty culling of paperwork for a major turning point in biological science.[161] That Darwin spent another year writing the *Origin of Species* about natural selection, after the reading of Wallace's "Ternate Paper," was also strange if his work was nearly complete.

Darwin's ideas were completely unpublished in 1858 and may have stayed that way forever had he not received Wallace's "Ternate Paper." (Note that the complete presentation to the Linnean Society is available online free of charge.[162]) Darwin and his friends chose the Linnean Society to present both Wallace's detailed letter and Darwin's writing about evolution by natural selection for a reason: the Linnean Society proved opportunistic for Darwin.[163] The members of the Linnean Society included Darwin's friends Thomas Bell (then-president of the society), Lyell (a member of the society and a friend of Bell), and Hooker (who "virtually ran the journal and saw the programme secretary constantly"[164]). Janet Browne writes in her book, *The Power of Place*:

> First and foremost, Wallace did not know anything about the proposal [of Darwin claiming dual priority and using Wallace's paper at the Linnean Society]. His [Wallace's] private communications to Darwin on a natural history matter, sent out to Lyell for comment, was to be announced without his [Wallace's] knowledge and as an accompaniment to writings about which he [Wallace] knew nothing. On the face of it, it looked as if Lyell and Hooker were suggesting that their friend Darwin—a man at the heart of scientific society—should not lose out to an interloper.[165]

Science, it seems, took a backseat to the politics. None of this was mentioned in the American Museum of Natural History's Darwin Exhibition. The order of the reading to the society placed Darwin first and Wallace second, an order that gave an impression, intended or unintended. If Darwin had completed his work on evolution by natural selection, or even had it close to complete, he would have been in a position to make it public for years. He had the financial means, knowledge of how to publish, personal contacts, social influence, and famous grandfather who published about naturalistic evolution. But Darwin did not publish his work on natural selection. Why? For twenty years, Charles Darwin had struggled unsuccessfully to formulate a complete explanation for evolution, but had not published so much as a line on the subject.[166]

Did Darwin not publish because he did not have his ideas about natural selection completed or settled in his mind or nearly ready for publication? Spurred by reading Wallace's "Ternate Paper," and in fear of being forgotten to history, Darwin made the "Ternate Paper" public at the Linnean Society—after reading the detailed material on evolution by natural selection that Wallace had sent to him.

Wallace's "Ternate Paper" was sent to the Linnean Society on June 30, 1858, along with Darwin's handwritten copy of a letter sent in 1857 to botanist Asa Gray, and the early sketch on evolution he completed in 1844.[167] Wallace's essay needed no further attention.[168] The same could not be said for Darwin's contribution,[169] which was not complete. Hooker and Lyell subtly justified their friend Darwin's position and, by implication, their own. Wallace's work was praised and then delicately pushed aside—merely the stimulus, they hinted, that encouraged Darwin to make a preliminary announcement. As Janet Browne writes:

> The material that followed reinforced that impression. [That is, the impression that Darwin was the main attraction. There was no hint of Wallace not knowing what was taking place.] The articles were arranged in alphabetic order by author, as was customary at the Linnean Society for double contributions. On this occasion, the alphabet coincided impressively with chronology. Darwin's extracts came first (in alphabetic order) and from and at the end, Wallace's complete February 1858 essay ["Ternate Paper"]. Darwin's priority reverberated from every page. Even Darwin winced when he saw the layout some weeks later. He assumed that his remarks would appear as a kind of appendix or as footnotes to Wallace. Privately embarrassed, he was relieved he had not personally supervised this printed reversal of fortunes.[170]

Darwin told Lyell, "I do not think that Wallace can think my conduct unfair, in allowing you and Hooker to do whatever you thought fair."[171]

Wallace was not in the same economic and social class as Darwin, but that does not excuse the wrong done to him. Wallace was ready to publish ahead of Darwin. In his 1980 book, *A Delicate Arrangement*, Arnold C. Brackman writes:

> Darwin notwithstanding, Wallace was the first to develop a fully coherent thesis to explain evolution [by natural selection]. His impact on history is incalculable. The fascinating, unanswerable question—raised by Darwin's intimate friends themselves—is whether Darwin would ever have written *Origin of Species* had there never been a Wallace.[172]

If Wallace had sent his paper to someone else, would Darwin have published anything? Would Wallace have history's position? Brackman's book, *A Delicate Arrangement,* tells of a "great wrong" in history of evolutionary work.

> The greatest sin a man of science can commit is to use another scientist's idea as his own. It tells of how Darwin and his two friends conspired to secure priority and credit for that theory for Charles Darwin.[173]

Brackman's book about Wallace and Darwin reads like a true-life detective story. By any reasonable account, Wallace preceded Darwin in going public. Darwin and his friends, upon Darwin receiving Wallace's letter detailing evolution by natural selection, made it appear that both were publishing together, and Darwin was in the lead. By the combined efforts of Lyell, Bell, Hooker, and Darwin, Wallace had his priority taken from him and turned into a joint venture of recognition for priority. Wallace earned and deserved priority before Darwin, which would make Wallace fifth for discovering natural selection. Darwin was then sixth in "discovering" and writing about natural selection, but that is not the impression one gets from American Museum visits and books written about Darwin.

Beagle Voyage

Darwin did not start his journey on the *Beagle* as someone who had no knowledge of evolution or reference material about it. He was prepared for the journey with a background given to him by his university teachers. He also had material available that was provided to him:

> When Darwin started the five-year journey on the *Beagle* as the paid companion to the ship's captain (not the ship's naturalist), he had a great start about the topic of evolution. The student Charles Darwin met two teachers that introduced him to evolution, then called *transmutation*. They were Robert Grant and Robert Jameson. Grant developed Lamarck's and Charles Darwin's grandfather's ideas of transmutation [evolution], investigating homology to prove common descent. Darwin also studied geology under Professor Robert Jameson whose journal published an anonymous paper in 1826 praising "Mr. Lamarck" for explaining how the higher animals had "evolved" from the "simplest worms"—this was the first use of the word "evolved" in a modern sense. Jameson's course closed with lectures on the *"Origin of the Species of Animals."*[174]

Was Darwin's 1831 Beagle voyage one of discovery or a voyage with evolutionary ideas already given to him by others? Darwin had taken material that he carried with him: evolution by Lamarck's "use and disuse," an encyclopedia that discussed island biogeography, and Lyell's material about evolution and geology. Darwin received mail during the trip before he landed at the Galapagos, providing him with material that helped him build a worldview and prepared him to filter and organize some of his observations to come. When did Darwin derive natural selection during his visit to the Galapagos? He did not. He derived it as an intuition. Australian Hiram Caton,[175] professor of politics and history at Griffith University, Brisbane, Australia, until his retirement, writes:

And natural selection? As Darwin tells the story, he didn't derive it as an induction from the Galapagos or other evidence; it came to him as an intuition, or better, a vision of living nature. He needed another two decades to assemble evidence.[176]

After receiving Wallace's Ternate paper in 1858, which spelled out natural selection, his friends hurriedly brought Darwin's scant notes and Wallace's completed paper, without Wallace knowing it. Darwin published the *Origin of Species* about one year later.

> As he [Darwin] was writing his classic, he learned, to his dismay [in 1858], that the young naturalist Alfred Russel Wallace had hit upon what he judged to be exactly his own prized concept [natural selection]. Uncertain what to do, he passed the challenge to friends [Lyell, Bell], who resolved credit for priority of discovery in his favour.[177]

3. Darwin Was Not First with Survival of the Fittest

In the opening of the presentation to the Linnean Society, Darwin had written:

> De Candolle, in an eloquent passage has declared that all nature is at war, one organism with another, or with external nature[178] [struggle for survival, survival of the fittest].

Darwin recognized the Swiss de Candolle as using survival of the fittest, struggle for existence, struggle for life, and competition. The poetic metaphor of "survival of the fittest" has taken root in culture and become a mantra, despite having no causal or operative meaning in nature or science: the phrases are emotive, not factual. At the very end of his 1859 *Origin of Species*, Darwin also uses de Candolle's war metaphor in his book, *Origin of Species*. At the close of the *Origin of Species*, Darwin writes:

Thus, from the war of nature, from famine and death, the most exalted object which we are capable of conceiving, namely, the production of the higher animals, directly follows.[179]

Applied to people, this metaphor would be a terrifying model (called social Darwinism when it is being applied to people, but in no way different than with animals). Its metaphoric terror is not causal and not science, but it does stir the emotions and minds of people. Likely not knowing of Wells' priority with the use of "survival of the fittest," Darwin adopted Spencer's "survival of the fittest" term in the fifth (1869) and sixth (1872) editions of *Origin of Species*, where Darwin writes:

I have called this principle, by which each slight variation, if useful, is preserved, by the term Natural Selection, in order to mark its relation to man's power of selection. But the expression often used by Mr. Herbert Spencer of the Survival of the fittest is more accurate, and is sometimes equally convenient...[180]

Spencer's use of "survival of the fittest" was first used in politics and is a grammatical tool applied to nature as one applies the terms "cold" and "hot," "big" and "small," or "near" and "far," none of which exists in nature, either. In 1831, Patrick Matthew uses several synonyms for the term "fittest" in his writing as follows:

[T]he self-regulating adaptive disposition of organised life may, in part, be traced to the extreme fecundity of nature... much beyond...what is necessary to fill up the vacancies caused by senile decay, as the field of existence is limited and preoccupied [now said to have "niches" occupied], it is only the hardier, more robust, better suited to circumstance individuals [fittest], who are able to struggle forward to maturity, these inhabiting only the situations to which they have superior adaptation and greater power of occupancy than any other kind; the weaker and less circumstance-suited being prematurely destroyed. this principle is in constant action; it regulates the colour, the figure, the capacities, and instincts; those individuals in each species whose colour and covering are best suited [fittest] to concealment or protection from enemies, or defense from inclemencies or vicissitudes of climate, whose figure is best accommodated to health, strength, defense, and support; whose capacities and instincts can best regulate the physical energies to self-advantage according to circumstances—in such immense waste of primary and youthful life those only come forward to maturity from the strict ordeal by which nature tests their adaptation to her standard of perfection and fitness to continue their kind by reproduction.[181]

In a February 22, 2006, review of Pietro Corsi's book, *The Age of Lamarck: Evolutionary Theories in France 1790–1830,* Australian reviewer Hiram Caton (Brisbane, Australia) writes about Julien-Joseph Virey and other French naturalists who preceded the English naturalists with many of the concepts of evolution, showing priority for many ideas belong to the French:

Corsi's book made me change the way I teach the history of evolutionary theory. I had accepted the Darwin-centric account, which says that modern theory begins with Charles Lyell and Darwin. Yes, Lamarck originated 'transmutation' of species thinking and Cuvier was an outstanding paleontologist. But Lamarckian inheritance doesn't work and Cuvier was a bitter opponent of evolution.

Thanks to Corsi's painstaking research we know that English evolutionary thought was time-lagged about a half century behind the French. The uniformitarianism vs catastrophism interpretation of earth history, which I had thought was due primarily to Lyell, was intensively debated by French geologists by 1800. The geologist Philippe Bertrand, proposed, in 1797, the marine origin of life and gradual evolution of all organic forms. Terrestrial plants and animals are descended from original marine species. Julien-Joseph Virey proposed (1816) that the term "evolution" be used to denote the transmutation of species. "It is thus plausible that, thanks to such evolution, nature has risen from the most tenuous mold to the majestic cedar, to the gigantic pine, just as it has advanced from microscopic animals up to man, king and dominator of all beings." In his Histoire naturelle du genre humain (1800) he stated the principle of sexual selection, which assured the optimum adaptive state through elimination of the weaker... We seem to hear Darwin speaking when Virey writes: "Nature initially produced only one very simple plant and one very simple animal, which it then varied to infinity, with gradual increases in complexity, to produce the most consummate species." The geologist Louis-Constant Prévost proposed that the evolutionary descent of each organism might one day be traced from the fossil record, from "the creation of the simplest beings to that of man himself."[182]

Quite apparently, as Hiram Caton points out, the French had their "Darwins" before the English did.

4. Darwin Was Not First to Use the Animal Breeders Analogy

Darwin had used the analogy of the animal breeders in Origin of Species as the basis for "selection" existing in nature. In addition to serving as an analogy, selection serves as a metaphor. The analogy, despite its invalidity of portraying of how nature operated during creation, was proposed as a model of how nature operated in the creation of new creatures. Analogies are not causal and cannot show how creation takes place. Darwin had a habit of personifying the term "nature." When dealing with the creation of new creatures, personifying nature deifies it. The deification results from having a name, "nature," and claiming that it created life, when all that really happened was that one deity was replaced with another. Using the term "nature" is little different than using a giant "X" for creation; it has no means of showing how creation took place: it is not causal. Further, Darwin often gave nature a person's characteristics, further amplifying his deification of the term. Without a model that showed which parts of nature created each of the body's many parts, together with the rules of creation and the relationships of how each part of the model operated with all the other parts, Darwin's definition of nature proved open-ended, allowing multitudes of options that made the term a slave to each person's views, rather than cementing it firmly to nature and making it mean the same thing to everyone. When he writes about nature and its actions, he may as well be writing about a deity, with many parts and capabilities under one name doing whatever he claimed. He defines nature as follows:

> So again it is difficult to avoid personifying the word Nature;
> but I mean by Nature, only the aggregate action and prod-
> uct of many natural laws, and by laws the sequence of events
> as ascertained by us. With a little familiarity such superficial
> objections will be forgotten.[183]

Darwin was wrong. The objections are not superficial. They are serious, legitimate, real, and still active, and they serve as beacons for Darwin's many failures. They are not forgotten and appear to become more intense with time. Breeders select animals for mating, and Darwin transferred that idea of breeders to nature selecting creatures as well. He didn't emphasize the differences between humans and nature, but those differences are fatal to his ideas. The breeders' intelligence, purpose, knowledge, values, plans, and goals would just as easily serve as an analogy for a supernatural intelligence, such as the God of Genesis. Even so, those qualities are not transferable to nature. When we know no more about the creation models after using them than before, as is the case with nature and natural selection, mysticism and deification become warranted charges. Nature certainly does not have the breeders' intelligence or other human characteristics, and the analogy fails. Darwin never showed how nature operated without the human characteristics. He only gave testimony that it operated successfully, a statement that is not science. In addition, the analogy of nature and animal breeders has no relation to nature or science.

Edward Blyth (1810–1873) was born in London and worked as a zoologist and pharmacist. He was one of the founders of zoology in India, and he corresponded with Darwin. He was recognized as the father of Indian ornithology. Blyth, like Wallace, refuted the use of an analogy of animal breeders. Blyth wrote three articles on variation, discussing the effects of artificial selection and describing the process in nature (later called natural selection) as restoring organisms in the wild to their archetype or normal body form rather than forming new species. These articles were published in the *Magazine of Natural History* between 1835 and 1837.[184] However, Blyth never actually used the term "natural selection." He wrote that any analogy was incorrect and did not portray what actually took place in nature. Here, two capable men used the same analogy conflictingly, providing an example of the failure of any analogy to represent what takes place in nature in a "cause and effect" model. The interesting thing about breeds, as Blyth writes, is as follows:

> If man did not keep up these breeds by regulating the sexual intercourse, they would all naturally soon revert to the original type.[185]

Wallace also disagreed with Darwin about the use of animal breeders and wrote that the breeders kept animals from reverting back to what they once were. Alfred Russel Wallace highlighted the same difficulty of using an analogy for selection and evolution as did Blyth. He writes much the same thing as Blyth about animal breeding in his paper titled "On the Tendency of Varieties to Depart Indefinitely from the Original Type" (called the "Ternate Paper").

Domestic varieties, when turned wild, must return to something near the type of the original wild stock, or become altogether extinct.[186]

Near the ending of his seminal 1858 Ternate paper, Wallace writes:

We believe we have now shown that there is a tendency in nature to the continued progression of certain classes of varieties further and further from the original type—a progression to which there appears no reason to assign any definite limits—and that the same principle that produces this result as a state of nature will also explain why domestic varieties have a tendency to revert to the original type.[187]

Long before Darwin used the animal breeders' analogy, Wells, the second person to use "natural selection," made the same claim that nature does what the animal breeders do. Wells observed, before the Royal Society of London in 1813, that all animals tend to vary in some degree, and that agriculturists improve their domesticated animals by selection. What is done in this latter case by animal breeding, he said, "by art, seems to be done with equal efficacy, though more slowly, by nature."[188] Darwin was not the first to use an analogy of animal breeders to portray natural selection, but he was one of a number of such claimants.

Summary

Most (if not all) of the major ideas credited to Darwin were actually discussed in print by others before him. Others preceded Darwin's use of evolutionary ideas, such as naturalistic evolution, descent with modification (differential reproduction), chance creation of variations, direction, gradualism, sexual selection, common ancestor, genetics, spontaneous generation's disproof, tree of life, denying special creation, theistic evolution, use of the term "evolution," ape-to-man transition, and homology. The Darwin Exhibition created by the American Museum of Natural History of the bicentennial of

Darwin's birth did not mention any of this. The industry supporting Darwin has created a myth surrounding him. It can only legitimately say that Darwin may have made natural selection popular, but this is a marketing accomplishment, not one of biology or science. Critics have even concluded that Darwin did not make any major new contributions to the theory of evolution by natural selection. It is easy to see why that charge is made, shown by the many concepts that preceded Darwin and are outlined or mentioned in this chapter. Many have concluded that Darwin borrowed all of his major ideas without giving due credit to others. Even the term "evolution" had gradually come to be used before Darwin used it. Darwin did not bring one new idea to evolution.

While not using the term "evolution," Robert Chambers popularized the subject with his book, *Vestiges*. Thomas Sims, who married Wallace's sister, felt that in *Origin of Species*, Darwin slighted Wallace by not giving due credit to him. Four years before Darwin had published *Origin of Species*, Wallace had written the February 1855 publication of his Sarawak law, titled, "On the Law Which Has Regulated the Introduction of New Species." Up to that time, Darwin had published nothing at all on evolution or natural selection. Darwin had used the benefit of a groundbreaking paper that Wallace sent to him from the island of Ternate on March 9, 1858,[189] containing the first complete exposition, in writing, of descent and divergence with modification through natural selection, which is now referred to as the "Darwinian" theory of evolution by natural selection.[190] Darwin had received Wallace's "Ternate Paper" in May 1858. Again, Darwin had published nothing on evolution or natural selection up to that time. Without Wallace's knowledge or approval, Darwin and his friends presented Wallace's complete "Ternate Paper" along with Darwin's unpublished excerpts from an incomplete sketch in 1844, which was then an unpublished fragment from the Asa Gray letter that Darwin had sent and kept in his extensive files. Darwin did not publish any work on evolution until after receiving Wallace's "Ternate Paper." Darwin hurriedly published *Origin of Species* within a year of receiving Wallace's completed paper and without acknowledging Wallace's work. This is indicative of Darwin's character, which may deserve to be called "self-involved."

Darwin "slight" to Wallace, which Sims charged was severe, was a self-serving aggrandizement at Wallace's expense. In the first four paragraphs of the first edition of the *Origin of Species*, Darwin used the words "I," "me," and "my" forty-three times. Darwin mentions Wallace's important 1855 "Sarawak Law" paper as a side note and does not mention Wallace's 1858 "Ternate Paper" in those paragraphs. He even omitted mention of them in the entire 491page *Origin of Species*. Darwin mentioned Wallace four times in the 1859 *Origin of Species* as if in passing, and with Wallace's work as a companion to his own work, again showing his self-involved character and unwillingness to give due credit to others.

Wallace's brother-in-law, Sims, rightly felt that Wallace did not receive due credit from-Darwin. The missing credit related to priority. In effect Darwin, through support from his friends, kept as much credit as possible for himself. Wallace, then unknown, poor, and a self-educated naturalist and adventurer, always acted the gentlemen. Darwin received the honor of being buried in Westminster Abbey because of the claimed importance of his discovery. It would appear that a great co-discoverer of natural selection, such as Alfred Russel Wallace, who published about evolution before Darwin and who sent his "Ternate Paper" to Darwin before Darwin ever published a single world about evolution, should also be buried in Westminster Abbey. The question naturally arises: why wasn't he?

CHAPTER 4

THEISTIC EVOLUTION

Introduction

Theistic evolution is the belief that God did not directly create new creatures as described in Genesis; rather, it holds that God created laws that were responsible for creation of all new creatures. Theistic evolution is not a scientific theory and does not have the characteristics of science models; it is a religious belief system about how evolution relates to religion. It is a perspective presented by the NAS in their booklet *The View* that takes "evolution by God" and changes it to "evolution by natural selection." It is dramatically different than the Christian and Jewish belief in Genesis. Theistic evolutionist supporters don't think of themselves as having any conflict with religion, when in fact they conflict significantly. They support the belief that religious teachings about creation and evolution by natural selection need not contradict each other, when in fact they do. The conflict arises with natural selection being held as creating new forms of life, which replaces God as the creator of new creatures: God is absent from creation, as is the case with atheism.

As this view has grave inconsistencies, it is held as spurious by some, not being what it purports to be. Those holding this view are sometimes described as Christian Darwinists.[191] An organization named BioLogos supports theistic evolution,[192] along with others, such as the National Academy of Sciences (NAS). According to theistic evolution, God was not present during creation of any creatures, including the creatures in the Garden of Eden, especially Adam and Eve. For theistic evolutionists, natural selection created all new creatures, including humans: God never created any form of life and never interacted with humans. With theistic evolution, God doesn't exist in any personal or communicational sense with people. This contradicts the Bible and those holding the biblical account to be correct. The NAS Booklet, *A View*, defines theistic evolution as follows:

> Many religious persons, including many scientists, hold that God created the universe and the various processes driving physical and biological evolution[193] and that these processes then resulted in the creation of galaxies, our solar system, and life on Earth. This belief, which sometimes is termed "theistic evolution," is not in disagreement with scientific explanations of evolution.[194]

Read the quote carefully or you may miss their point. It is that many "religious persons" and "scientists" believe that God created processes driving biological evolution but did not create living creatures. In the NAS's view, God did not create life, animals, or people, which is a religious view. God was never active in direct creation of life.

The NAS presents their view in that quote. Is it the NAS intent to have people of other religions adopt those views? Contrast this claim with the section about "special creation" in chapter 2, "Models of Evolution's Cause." The comparison is blindingly stark. God's role as portrayed here by the NAS and in Genesis in the special creation section are complete contradictions of each other. One would expect that a religious person would not agree with the NAS's position about special creation by God. Yet, "religious persons" are cited by the NAS as agreeing with natural selection being responsible for creation of new creatures. Why? Who are the "scientists" in the NAS quote? What special credentials do they hold that they should cited by the NAS? Do they belong to the Christian, Jewish, Muslim, atheist, or Darwinist religions?

The above NAS quote actually describes a religious belief system. It is not science. Beliefs about God have never been shown to operate in nature as science. Beliefs do not constitute science, and they are not found in nature's physical world. One may wonder why they are discussing others' religions, such as the Young Earth Christians and, indirectly, all Christians. Newton kept his Christian religious views out of the science models. The science models he discovered cannot reference beliefs because they would then not be operative in nature. All the major religions deal with four significant topics, as does the above NAS quote:

- Origin: How did life originate? (Where did people come from?)
- Purpose: What is man's purpose? (Why are people here?)
- Morality: What are the rules people should live by?
- Destiny: What happens to each person after death?

Any system of thought, especially one of "cause and effect" that renders answers to these four topics, is identified as a religion. The NAS "religious person" subscribes to a system of thought that answers these four topics, regardless of what that religion is called. Natural selection answers these four points, as does the religion of atheism. The religion of Darwinism does as well.

Two conflicting worldviews, such as Darwinism and Christianity, each channel very different answers to the four preceding points. Natural selection provides answers to the four fundamental topics testimonially filtered through natural selection. The NAS booklet, *A View,* defines a "religious person" and gives the impression that such a person may possibly believe in God, the Bible, and natural selection—and still be a Christian. But its authors do not elaborate on that claim. As the NAS tells us, theistic evolution beliefs are not in disagreement with scientific explanations of evolution, and that a religious person can subscribe to them; as a result, one may wonder if the NAS believes that God can be discussed in science and biology classrooms. They remain silent on this point. It is consistent with their claim, and they should have discussed it. As they tell us, the NAS's "religious person"[195] is a theistic evolutionist.[196] Can such a person (who is to maintain a consistent set of beliefs) be Christian, Jewish, or Muslim and also be a theistic evolutionist? Part of the answer comes from Thomas Huxley, who wrote about religion in 1860. Was it true then? Is it true today? Huxley writes:

A deep sense of religion is compatible with the entire absence of theology.[197]

In other words, one does not need God to be "religious." One can be an atheist or an agnostic (as Thomas Huxley claimed himself to be). If this is what the NAS meant by a "religious person," it was between the written lines of their booklet and was tacitly implied, but not stated. Huxley should have known that worshiping a creator is dramatically different than worshiping with a deep sense of the "absence of theology." Worshiping the Judeo-Christian God is different than having "a deep sense of religion" about one's work or about dedication to the power of the state. Is being a secularist or atheism what the NAS meant by a "religious person"?[198]

Quite apparently, the NAS thinks that one doesn't need God to be a "religious person."[199] With natural selection doing all the creating, God is without any presence. It follows that God also did not give any rules of life (morality) to live by. Huxley gives a definition of religion indirectly when he writes of himself:

> Most of my colleagues were "ists" [socialists, atheists, etc.] of one sort or another...so I took thought, and invented what I conceived to be the appropriate title of "agnostic."[200]

Agnosticism, for the group Huxley refers to, was a religion of skeptics.[201] Most, if not all, practiced atheism. By coining the term "agnosticism," Huxley was labeled "Pope Huxley" in the January 29, 1870, issue of *Spectator*. He had coined his own religion, essentially forming one about natural selection. He was a "religious person" and did not have or need a God to believe in.

A religion need not be about God or a supernatural force, and its definition should acknowledge this fact. A religion may be about the natural (atheism) or supernatural (theism). Webster defines religion as relating to or manifesting faithful devotion to an acknowledged ultimate reality or deity.[202] Wikipedia defines a religion as an organized approach to human spirituality that usually encompasses a set of narratives, symbols, beliefs, and practices, often with a supernatural or transcendent quality, that gives meaning to the practitioner's experiences of life through reference to a higher power or truth.[203]

When scanning the history of evolutionary models, no evolutionary creation model operates in nature, and not one is based in science, necessitating the use of testimony as the link between creation and evolution. Natural selection clearly contains nothing of nature's parts or processes, making it impossible to operate in nature or be considered science. Natural selection fits the definition of a secular spirit: empty of nature but identified and attributed with creation capabilities. It is a deity by a different name; in fact, it is a trinity called "chance, selection, and direction." Although the "chance" in natural selection is easily misrecognized as being part of nature, not one component of the trinity is ever defined using nature's components. There is no physical population to any one of them, which would then define and fix nature's components that are being used. Gaming halls, in contrast, have a fixed population for each game, such as fifty-two cards in a deck or six dots on each die in a game of dice.

Natural Selection and Morality

A religion is also a source of morality that sets permissible rules for behavior, such as the Ten Commandments given by God to Moses. Robert Winthrop, a founding father of the United States of America, says it succinctly and accurately in 1849 when he discusses morality:

> All societies of men must be governed in some way or other. The less they may have of stringent State Government, the more they must have of individual self-government. The less they rely on public law or physical force, the more they must rely on private moral restraint.

> Men, in a word, must necessarily be controlled, either by a power within them, or by a power without them; either by the Word of God, or by the strong arm of man; either by the Bible, or by the bayonet.[204]

Darwinism, atheism, and theistic evolution all provide a morality without reference to a God, but it is a morality that is radically different than that commonly accepted as being from God. The source of morality without the God of Genesis was discussed during the famous 1948 BBC radio debate on the existence of God between Father Copleston and atheist Bertrand Russell. During that debate, the moral argument was introduced. This debate illustrated how atheists such as Bertrand Russel determine morality. It applies to Darwinism and those who subscribe to natural selection to this day, including theistic evolutionists. The famous BBC Radio debate showed Russell describing how he determines morality without the biblical God. The contents of the Russell and Copleston debate are as applicable then as today:

> **Father Copleston:** [W]hat's your justification for distinguishing between good and bad or how do you view the distinction between them?

> **Bertrand Russell:** I don't have any justification any more than I have when I distinguish between blue and yellow. What is my justification for distinguishing between blue and yellow? I can see they are different.

> **Father Copleston:** Well, that is an excellent justification, I agree. You distinguish blue and yellow by seeing them, so you distinguish good and bad by what faculty?

> **Bertrand Russell:** By my feelings.

> **Father Copleston:** By your feelings. Well, that's what I was asking. You think that good and evil have reference simply to feeling?[205]

With no absolute source of morality, such as from the God of Genesis, and no absolute rules for how to live, no absolute morality can exist. Only feelings are used to form judgments, with hedonism, sensualism, debauchery, and genocide being on the menu. With natural selection, where the biblical God is absent from the physical world and has no relationship with humans, all rules for how to live come from feelings, as Bertrand Russell showed, which is called "relativism" (also called "arbitrariness"). Some people call this "social Darwinism," which is survival of the fittest applied to animals and people (who are merely animals that were created by natural selection after the others). Robert Winthrop recognizes the failures of a government without religion to constrain its actions when he writes:

> It may do for other countries and other governments to talk about the State supporting religion. Here, under our own free institutions, it is Religion which must support the State.[206]

Of the fifty-six men who signed the Declaration of Independence, nearly half held seminary or Bible school degrees.[207] Other founding fathers echoed the same views, with over 95 percent being practicing Christians.[208] Their morality was from an absolute source that they used to establish and govern a nation. It was Benjamin Franklin who addressed the Constitutional Congress regarding prayer to the Almighty being needed before their meetings. Franklin called them to prayer on June 28, 1787. He addressed the group with the following words:

> I have lived, Sir, a long time, and the longer I live, the more convincing proofs I see of this truth—that God governs in the affairs of men. And if a sparrow cannot fall to the ground without His notice, is it probable that an empire can rise without His aid?[209]

> We have been assured, Sir, in the Sacred Writings, that "except the Lord build the House, they labor in vain that build it." I firmly believe this; and I also believe that without his concurring aid we shall succeed in this political building no better

than the Builders of Babel...I therefore beg leave to move—
that henceforth prayers imploring the assistance of Heaven,
and its blessing on our deliberations, be held in this Assembly
every morning before we proceed to business, and that one or
more of the clergy of this city be requested to officiate in that
service.[210]

Benjamin Franklin echoed the other founders of the United States. Darwin worked to remove that creator. Over the past century, one naturalistic theory has gained the most prominence: evolution by natural selection. The goal of natural selection's supporters, including theistic Christians, is to replace evolution by special creation. Darwin declared God to be removed from creation as his goal in the *Descent of Man*, when writing, "[I] hope, done good service in aiding to overthrow the dogma of separate creations."[211]

It is a self-contradiction to try to develop a scientific model to compete with religion as the models with the characteristics of science always exist independently of all religions: no model that is science conflicts with religion. For example, Newton and Pasteur believed in God, but their models (based in nature and science) remained independent of their views, as well as of all morality. Darwinists have redefined the source for the four fundamental points of religion as coming from natural selection; with that view, morality does not come from God but rather from one's feelings. God was removed for these people, and they thought they replaced Him. In their worldview, each person creates his or her own rules, ethics, and morals. For Darwin, God did not set the rules of behavior. Natural selection provided the basis for all rules. This is Darwin's "far more satisfactory morality," of which he writes in the 1859 *Origin of Species*:

Finally, it may not be a logical deduction, but to my imagination it is far more satisfactory to look at such instincts as the young cuckoo ejecting its foster brothers [killing them],—ants making slaves,—the larvæ of ichneumonidæ feeding within the live bodies of caterpillars,—not as specially endowed or created instincts [by special creation], but as small consequences of one general law [natural selection], leading to the advancement of all organic beings, namely, multiply, vary, let the strongest live and the weakest die. We may console ourselves with the full belief that the war of nature is not incessant, that no fear is felt, that death is generally prompt, and that the vigorous, the healthy, and the happy survive and multiply.[212]

It is not the model of natural selection that shows nature's operations in the quote; it is Darwin. Natural selection is inoperative in nature, whereas Darwin is describing what it is doing. This is the opposite of the science models of Newton, Copernicus, Pasteur, and Mendel. In the above quote, Darwin elevates himself above God when he describes his morality as being "far more satisfactory."[213] In the above quote, Darwin presents a way to "think about and live with nature," including humanity's role. Darwin's God was indifferent to everything that was to take place. In every practical way, the God of Darwin (and the God of the NAS) remains uninvolved with the creation of new creatures, including humans. In every practical way, with natural selection as the creation model, there is no God involved with humans. God is an unnecessary "add-on" to natural selection, for no faith system can possess the characteristics of science, especially the faith surrounding natural selection. With natural selection, the religion of atheism prevails over all other religious claims.

Theistic Evolution, Creation, Darwin, and the NAS

In direct conflict with Darwin, Wallace thought that God and natural selection were both involved with creation in some way. Darwin's American friend, Asa Gray, also thought God was directly involved with natural selection and creation as well. Each of them used natural selection, and each had different views of how it operated. This difference of natural selection's operational visions would be difficult to reconcile, even impossible, if the model were dependent on nature and operated as science, but it does not operate as science, or each would have the same identical answer. If it is said that God created natural selection and the first creatures on earth, it can legitimately be asked, "How does one determined that creation by God started, was performed one or more times, and then stopped?" Even without waiting for a response, it can be seen that faith was used to claim that God created natural selection and then the first creatures. Science would not be involved, but religion would be. This would be the faith of Darwin, Dobzhansky, and, because of their endorsement of the "religious person," the NAS. Claiming that God created life only once seems to imply that Darwin had such a vast knowledge about God and his practices that he knew as much as God. Knowing anything less would mean that Darwin did not know what God was doing, or how, or why, and could not legitimately make the claims he offered to others. The same holds true for Dobzhansky and the NAS.

Darwin's God is the God in theistic evolution. Darwin defined chance in natural selection (a "chance" without any population in it) in such a way that it does not even allow Darwin's God to know what will take place next in creation, making God less than all-knowing, less than all powerful, and less than perfect.[214] If God were to know the outcomes of such a chance, he would bear responsibility for the evil that Darwin imagined existing in the world. In Darwin's mind, God needed to be saved from any association with evil, and Darwin thought he was the person to save God.

Wallace, the cofounder of natural selection, believed that natural selection created all creatures' bodies, including some (but not all) of humans' bodies. Wallace believed and tested his theory that only God could have created humans' "higher qualities," such as the "higher feelings" of pure morality and refined emotion, and the power of abstract reasoning and ideal conception."[215] In conflict with Darwin, but using the same model of creation, Wallace felt that only a higher power could have created these qualities in humans. Wallace writes:

> If, therefore, we have traced one force, however minute, to an origin in our own will, while we have no knowledge of any other primary cause of force, it does not seem an improbable conclusion that all force may be will-force; and thus, that the

whole universe is not merely dependent on, but actually *is*, the will of higher intelligences or of one Supreme Intelligence.[216]

Wallace did not believe in the God of Genesis. He maintains, "Whether we call it God, or spirit," it must play an important role in human evolution.[217] Darwin did not agree with Wallace about his inclusion of God with natural selection any more than he did with Asa Gray, who thought God directed variations. Here we have different outcomes from the same model of creation—natural selection—demonstrating again that a testimonial model becomes what you testify it to be using a personal worldview. Creation models use testimony to portray the operations of nature, its rules, and its relationships, such as natural selection, are excluded from being science. As the beginning of the last chapter of the 1859 *Origin of Species* shows, Darwin's entire book is shown to be one long testimony (in the form of an argument), further showing that natural selection has no relationship to nature—except through Darwin's testimonial links. He writes:

> As this whole volume is one long argument, it may be convenient to the reader to have the leading facts and inferences briefly recapitulated.[218]

The *Origin of Species* is one long argument that is meant to convince others to accept Darwin's claims. Models based in science do not use arguments and do not require or use an individual's worldview. Only faith-based models hold that requirement for testimony, which is why Darwin provided over four hundred pages of it in his book. The NAS booklet, *A View*, does not mention Wallace's inclusion of God[219] in creation. The differences between the views of Wallace, Gray, Darwin, and others should be included in biology and science classrooms. There is no doubt that Wallace was a "religious person." Wallace thought his tests of natural selection were scientific, just as Darwin thought his were scientific, even though they contradicted each other. God existed in Wallace's view of natural selection when applied to the creation of humans. Wallace even successfully tested that view (see chapter 11, "Testing Miracles and Natural Selection"). Darwin rejected Genesis, wherein God has a special relationship with humanity. The NAS must agree with Darwin's non-Genesis view, as it states in the NAS booklet, *A View*:

> Scientists as well as educators have concluded that evolution—
> and only evolution [by natural selection]—should be taught
> in science classes because it is the only scientific explanation
> for why the universe is the way it is today.[220]

The NAS is incorrect. Evolution is only half of the causal model: the half that was created. It is not an "explanation," because it is only half of the creation model. The full creation model that Darwin claimed is "evolution through natural selection," showing natural selection as the "cause" and "evolution" as the part that was created. The NAS claim that "only evolution—should be taught in science classes"[221] omits the cause that should be taught. Evolution may be said to be caused by special creation. There may be many causes to evolution, but none of them is dependent on nature, and none of them is science. For its own reasons, which are not clarified in the booklet, the NAS uses the term "evolution" but not the accurate and complete term, "evolution by natural selection," which shows the effect (evolution) and the cause (natural selection). They do not explain the reason for this omission. Perhaps it is because natural selection is not science and cannot be successfully defended as a science model. As soon as the cause of evolution is included, a seemingly never-ending debate follows—a debate that Darwinists are losing. No one objects to the creatures that form evolution. No one objects to science being taught in the classroom. They object to natural selection, the icon of the religion called "Darwinism," being taught in the science and biology classrooms.

It Is the *Cause* That Is Debated

Evolution may have different causes, with each cause being determined by the use of different worldviews of different people. The cause of evolution is in contention, whether it is evolution by natural selection,[222] or evolution by special creation, or evolution by God's laws, or evolution by orthogenesis.

It is the cause that makes a great deal of difference to everyone, including Dobzhansky and the NAS. Darwin tried debunking Genesis and special creation as he thought it was a competing cause to natural selection. Darwin was not arguing about the creatures that make up evolution, just the cause of those creations, a cause with which he argued throughout his book. Omitting the cause and merely writing "evolution" instead of "evolution by natural selection" conceals the true nature of the debate. The debate is not about "evolution and religion" or about "science and religion," as is often portrayed. The actual debate focuses around natural selection as faith-based. It is about natural selection not being part of nature. It is about the need to constantly link cause and effect with testimony, as so many books about natural selection demonstrate. Many atheists and agnostics reject natural selection, along with Christians, Jews, and Muslims. Many people never agreed with natural selection as a creation model, nor special creation, nor theistic evolution. Many of Darwin's close friends, such as Thomas Huxley and Charles Lyell, disagreed with evolution by natural selection. They accepted evolution, but not natural selection. Even Spencer, who provided Darwin with "survival of the fittest," disagreed with natural selection, as he was a Lamarckian. The NAS booklet, *A View,* does not mention these people and their views. It should.

Darwin rejected evolution by special creation. By endorsing Darwin and accepting theistic evolution, it follows that the NAS rejects special creation as well. In direct contradiction to Darwin, Wallace agreed with evolution and God's involvement as a cause. In some cases, Wallace believed in "evolution by God." Wallace still called his view "science." He used "testimonial testing" and successfully tested God's involvement (see chapter 11, "Testing Miracles and Natural Selection"). Wallace and Darwin, among others, openly debated God's involvement in creation, with Wallace being selective about God's role and Darwin refusing God's role completely. We should ask, "Is the nature of *that* debate to be included in science or biology classrooms?" Is Wallace's successful testing of God's involvement with evolution to be presented in the classroom? It should be included as it is part of the "scientific process" as it is tested successfully. It is science, after all—as much science as Darwin's "testimonial tests" are science. The public debate about evolution focuses on the *cause* of creation, which is the cause of evolution; shouldn't that be included in the biology and science classrooms?

Age of the Earth—a Straw Man Argument?

The NAS grants importance to other people's religions, so much so that they include them in the title of the booklet, *A View*, when they referred to "Science and Creationism."[223] "Creationism" is a term often used to refer to a specific group of Christians with specific views on the age of the earth. Sometimes they are called "Young Earth Christians" or "Young Earth creationists." The Young Earth creationists are "religious persons," just not the religious persons the NAS approves of. The NAS tells us that these Christians hold the earth to be 6,000 to 10,000 years old, when today's estimate of the earth's age is 4.7 billion years. They do not tell us that natural selection does not contain any reference to time and offers no means of relating to the pace of creation. Natural selection does not contain billions, millions, or thousands of years. The age of the earth is computed *apart* from natural selection, which shows it to be inoperative in nature—again. The age of the earth, while important, is a straw man topic, possibly an intended diversion from more important topics they would rather avoid. There is no shortage of straw man topics to divert attention away from natural selection being a failure due to having no parts of nature in it, being inoperative in nature, and depending on groups such as the NAS testifying on its behalf, as is the case with miracles.

Another straw man argument by the NAS is this: "Some religious groups deny that microorganisms cause disease, but the science curriculum should not therefore be altered to reflect this belief."[224] Certainly, this statement is not part of evolution by natural selection and not part of the public debate. It should not have been mentioned, as it is more of a distraction and not a problem with evolution or natural selection. Another point introduced in the NAS booklet, *A View*, is "consensus" as a means of determining what is or is not science, which in fact is incorrect. Votes do not make the earth flat or round. They do not cause events to take place. They can, however, intimidate those who disagree with consensus; and perhaps that is the reason for its use.

The NAS writes, "The scientific consensus around evolution is overwhelming." This is in fact incorrect. Scientists have their religions and opinions, but none of them is science—and neither is consensus. Evolution exists in nature, but evolution through natural selection does not. In the quote by Fisher, at the beginning of this book's first chapter, he states, "Evolution is not natural selection." That is what the debate is about—the cause of evolution, not that it exists. For the NAS, that cause is natural selection. It may be that the NAS's strategy is to avoid giving natural selection separate consideration due to its being an abject operational failure in nature's world. By it having no physical components in the model, it could not be anything else but a failure. Having a testimonial link between cause and effect, the model of natural selection will always remain a matter of faith, not science.

Many topics prove more important and more urgently in need of debate concerning natural selection, such as natural selection's lack of nature's physical world in it, making it incapable of representing that physical world. For instance, a topic that needs discussion is "chance", claimed to create the "raw material" in natural selection's first of two parts. An interesting discussion would be the "populationless" nature of chance, leaving nothing physical by which it may be determined. A good discussion in the classrooms would be, "Without anything physical, does natural selection even exist?" The foundation of Darwin's model, "good variations", is undefined and never shown to exist in nature,[225] making that a discussion point that is more critical than "Young Earth Creationists." The NAS could elaborate upon how to determine the number of variations in an accumulation or even if such things exist, but they leave this topic unexplained and impossible to determine. An interesting point that needs the sunlight of nature's physical cause and effect is the manner by which bad variations are discarded, but this topic is left to the imagination.[226] More revelation is needed concerning the relationship between the accumulation of variations and new body parts that are created, but this is left as a simple "self-evident truth", one determined by testimonial fiat. A key discussion point could also be "selection,"[227] which is left as a metaphor, having no physical components to define its physical operations. Similarly, there are mysteries surrounding "direction", located in the second part of natural selection, which is no more than a metaphor, with the all too apparent image of it being a backdoor inclusion for "design" and goals, working mysteriously in conjunction with "selection." The NAS should write booklets about how they resolved these mystically appearing issues, allowing them to call them science, thus removing the mysticism that currently surrounds them. It would appear that the issues are impossible to resolve, disabling natural selection from portraying nature's creation of new body parts or new creatures. In place of discussing any or all of these important topics, the NAS focuses upon straw man issues such as microorganisms causing disease, the Young Earth creationists, and the age of the earth, which are important, but would disappear if the important issues were causally and not testimonially resolved. The NAS's concern about the age of the earth is stated as follows:

> The advocates of "creation science" hold a variety of viewpoints. Some claim that Earth and the universe are relatively young, perhaps only 6,000 to 10,000 years old.[228]

It needs to be pointed out that natural selection does not even have "time" represented in it. The age of the earth may be a topic in biology or science classroom, but it leads to the question, "Why focus on it?" Natural selection, without time represented in it, cannot be related the age of the earth except through testimony, but testimony is not science.

The science and biology classrooms should be discussing the legitimacy or illegitimacy of declaring that "evolution is a historical process"[229] and noting that "past stages cannot be observed directly thereby inferring answers to creation from that context."[230] This topic is critical to what is or is not science. For instance, students could discuss the fact that inference is not causal in nature and thus not science. The legitimacy of natural selection and all creation models is derived or destroyed by such discussions. Students could note that such discussions seem to always fall back on faith-based answers with intense worldviews–based influences. These classrooms discussions would be educational.

The 1925 Scopes Trial's (Monkey Trial) Christian view of the age of the earth would have been a great point of discussion. The biblical Christian William Jennings Bryan was notable in his time and the famous prosecuting attorney in the case. Bryan was a member of the American Academy for the Advancement of Science[231]. He had more degrees than the expert witnesses in support of Darwin's theory.[232] Yet, Bryan was is often not portrayed that way – Why? Bryan's official testimony at that trial showed that he believed strongly in Genesis and the Bible but not in any specific age for the earth. In the *Scopes trial* he testified about the earth's age under examination by Clarence Darrow, a lawyer known for defending labor leaders and radicals as well as high-profile murderers.[233] Darrow asked the sixty-five-year-old[234] about the earth's age during the trial.[235]

> From the 1925 Scopes Trial records (also known as the Scopes Monkey Trial):
>
> **Darrow**: Mr. Bryan, could you tell me how old the earth is?
> **Bryan**: No, sir, I couldn't.
>
> **Darrow**: Have you any idea how old the earth is?
> **Bryan**: No.
>
> **Darrow**: Do you think the earth was made in six days?
> **Bryan**: Not six days of twenty-four hours.
>
> **Darrow**: What do you consider it to be?
> **Bryan**: I have not attempted to explain it…
>
> **Darrow**: You think these were not literal days?
> **Bryan**: I do not think they were twenty-four-hour days.
>
> **Darrow**: What do you think about it?
> **Bryan**: That is my opinion—I do not know that my opinion is better on that subject than those who think it does.

Darrow: You do not think that?

Bryan: (Christian): No. But I think it would be just as easy for *the kind of God we believe in* to make the earth in six days as in six years or in six million years or in six hundred million years. I do not think it important whether we believe one or the other. [Italics added].

Darrow: Do you think those were literal days?

Bryan (Christian): My impression is they were periods, but I would not attempt to argue against anybody who wanted to believe in literal days.

As shown from the 1925 Scopes Trial transcripts above, William Jennings Bryan's view on the age of the earth represented the view of many Christians then, as now. As Bryan's testimony shows, as a Biblical Christian, he did not believe the earth to be 10,000 or 12,000 years old, as the Young Earth creationists believe. Bryan did not hold a firm position for the age of the earth. It was not important to him. A topic of discussion could be that natural selection does not contain *time* as one of its components that deals with dating, which is an important part of nature, leading to the question, "By what means is the rate of creation by natural selection determined?" The answer echoes back as it always does: by testimonial attribution. Earth ages are independent of natural selection.

The Young Earth creationists cited by the NAS differ from many other Christians, but they have in their ranks many PhDs in biology and science. When Darwinists debate Young Earth creationists, they often lose. But the NAS does not mention any of this in their booklet, *A View*. Not all Christians necessarily hold the same beliefs about the age of the earth, just as not all atheists believe in natural selection (such as Huxley, Lyell, Spencer, and Wallace in some cases), but the NAS does not tell you that in their booklet.

Misrepresentations of Bryan's view of the age of the earth were published in his day, as they are today. In 1923, in a published response, Bryan writes:

The only persons who talk about a twenty-four-hour day in this connection do so for the purpose of objecting to it; they build up a straw man to make the attack easier, as they do when they accuse orthodox Christians of denying the roundness of the earth, and the law of gravitation.[236]

As others have in the past, could the NAS have fabricated a straw man just to knock Christians down? To many, it would appear so. Could they have injected the discussion of some people rejecting microorganisms causing disease to have people stand back and take sides with the NAS? Again, it appears so.

Theistic Evolution, Dobzhansky, and the "Light of Evolution"

The NAS identified Theodosius Dobzhansky as a "religious person." He was one of the contributors to the modern synthesis, which helped revive natural selection. Dobzhansky immigrated to the United States in 1927 and worked with Thomas Hunt Morgan at Columbia University, who had pioneered the use of fruit flies in genetics experiments. Dobzhansky published one of the major works of the modern evolutionary synthesis in 1937. It synthesized evolutionary biology with genetics and is titled *Genetics and the Origin of Species.*[237]

Dobzhansky believed in "evolution by natural selection," as he writes in his March 1973 article in *The American Biology Teacher,* titled "Nothing in Biology Makes Sense Except in the Light of Evolution."[238] His views were deemed important enough to publish in a journal for K–16 biology teachers and significant enough to merit highlighting by the NAS in their booklet *A View.* Dobzhansky's "Light of Evolution"[239] article has no roots in science as it has no cause and effect involving nature. In his article presented to a teaching organization, he takes the role of a preacher, not someone championing the cause and effect of nature that is supposed to be science. He writes:

I am a creationist *and* an evolutionist. Evolution is God's, or Nature's method of creation. Creation is not an event that happened in 4004 BC; it is a process that began some 10 billion years ago and is still underway. [Italics in original][240]

Dobzhansky misuses the term "creationist" in the above quote, for he is not a creationist as the term is commonly used and as the NAS uses it in their booklet *A View*. By his claims, Dobzhansky rejects Genesis and opposes any view of the direct creation of life by a creator, which shows a religious difference between him and creationists, such as Biblical Christians or Young Earth creationists. He opposed direct creation (special creation) as portrayed in Genesis. Dobzhansky is not the creationist portrayed by the NAS in their booklet, *A View*. They don't criticize Dobzhansky but rather hold him up as a model to be emulated. Are there two standards of judgment that are used by the NAS? Two types of creationists? Dobzhansky also called himself an evolutionist and believed in biological evolution as formulated by Charles Darwin. For Dobzhansky, creation was an unending process by natural selection, not continuing separate instances of special creation. He does not make that readily clear, possibly confusing some of his audience. In Genesis, creation ceases on the seventh day, but Dobzhansky denied this, claiming that creation by natural selection is still "underway." By claiming that he was a creationist, he could only mean that he believed in God *creating* natural selection and then natural selection creating new living creatures, thereby making Dobzhansky a "creationist," at least in his own eyes.

Contrasting Dobzhansky's metaphor of "light of evolution" are other metaphoric "lights," such as the light of the world, the light of life, and the light of knowledge. Each metaphor signifies a way of looking at the world removed from nature and science. Each metaphor expresses a view but cannot mean anything scientific. As with all metaphors, each reveals a view, each uncovering something of importance to that person, something that reveals that person to others. In this case, the "light of evolution" metaphor revealed something important about Dobzhansky and the NAS: their values, beliefs, faith systems, temperaments, things held in high regard, and things disliked. It also revealed that they opposed special creation, a belief held by Christians. Does this opposition emphasize an NAS religious view? In Dobzhansky's article, God created natural selection, then left, never again to interact with humans. This view is not one of science, and it begs the question: would the American Biology Teacher (ABT), a peer reviewed professional teacher's journal, allow Christians to print their religious views and present them as they allowed Dobzhansky in printing his religious views? Are there two sets of criteria for who gets to write or speak? Citing natural selection seven times in his article, four of which are personifications (deifications of natural selection),[241] Dobzhansky writes:

> The organic diversity becomes, however, reasonable and understandable if the Creator has created the living world not by caprice but by evolution propelled by natural selection.
>
> Natural selection may cause a living species to respond to the challenge by adaptive genetic changes.
>
> Natural selection is at one and the same time a blind and creative process. Only a creative and blind process could produce, on the one hand, the tremendous biologic success that is the human species and, on the other, forms of adaptedness as narrow and as constraining as those of the overspecialized fungus, beetle, and flies mentioned above.
>
> Natural selection does not work according to a foreordained plan, and species are produced not because they are needed for some purpose but simply because there is an environmental opportunity and genetic wherewithal to make them possible.[242]

The second, third, and last quotes above are personifications that deify natural selection, just as Darwin deified natural selection about one hundred years earlier. Some call creation by the Creator "design," whereas Dobzhansky called it "caprice." How would he have had such personal knowledge about

God to determine if it is caprice, design, or a completely uninvolved God? Did Dobzhansky consider himself a prophet? In science, to determine how a model of nature operates when creating new body parts, one has to examine the model itself. It is the names of the many parts of nature in the model that are shown to interact with the creature in such a way as to cause specific new body parts to be created. There is supposed to be a cause and effect relation between the many parts of nature in the model and the creations to which they give rise. However, Dobzhansky does not show that there are any parts of nature in the model. He does not show that there are body parts in the model. He does not show that there are any relationships of any kind in the natural selection. He merely makes fiat declarations which others are supposed to accept.

A model that operates in nature shows each of the terms in the model found in nature. Thus, the relationship and rules used to create new creatures' body parts, starting from the parent creature, can be found in the model, not in someone's testimony. The parts of creation cited by Dobzhansky are not in natural selection, not in nature, and are absent altogether. Simply, natural selection does not contain the terms used in Dobzhansky's article, which include "diversity, purpose, and blind opportunity." The model does not contain "success, species, adaptedness, fungus, beetles, flies, plans, or genetics." Nor does it contain any way that those terms could be identified. Dobzhansky does not bridge those terms to the model to show how they relate to creation. If natural selection were creating new creatures, there would be no need for anyone to attribute or personify capabilities to it as Dobzhansky did.

We don't measure a model of creation or hold it to be science by how well someone thinks it competes against other models but rather by how it represents nature's operations when creating creatures. Relating that to the link of nature's physical cause to effect show how creation takes place shows natural selection is not active or operative, but Dobzhansky's testimony is very active in this case. Each creature is composed of an endless number of chemicals, but natural selection has no chemical names used in the body. The search for an explanation about creation should go no further than the model itself. Since chemicals are in each body, chemicals must be in the model that created those bodies. If the model does not contain those chemicals, then it has no way of causing a creature's creation and bodily operations. One might as well have the letter "X" represent the processes of creation by nature. Dobzhansky's mention of God does not show natural selection in any different "light of evolution": it remains inoperative and a failure in nature, a failed testimonial model of creation—as all others have been in nature.

The Light of Evolution

When Dobzhansky writes the phrase "light of evolution," he borrows it from the Jesuit priest Teilhard de Chardin, whom, it is reported, Dobzhansky much admired.[243] In the last paragraph of Dobzhansky's article,[244] he quotes Teilhard de Chardin as follows:

> Is evolution a theory, a system, or a hypothesis? It is much more it is a general postulate to which all theories, all hypotheses, all systems much henceforward bow and which they must satisfy in order to be thinkable and true. Evolution is a light which illuminates all facts, a trajectory which all lines of thought must follow this is what evolution is.[245]

This quote is unscientific and false. Theories do not "bow", as if before a monarch. As encountered so often with natural selection, this statement is faith-based, without any cause and effect that involves nature. Dobzhansky portrays evolution as the underlying quintessential core of all of life and creation. Reading this belief, it would seem that one could not understand anything in the physical world of nature without understanding evolution. The statement is a conferring of royalty on evolution by natural selection. Nowhere does the article justify this view, and it is a view without any support from physical nature, nature's processes, or science. The magnitude of the statement's error is exemplified by the fact that one could study science through a PhD and beyond in engineering, including bioengineering, and never need to know anything about evolution—by any cause. Evolution really is that irrelevant to science and the operation of nature. The cause of evolution is of such broad unifying magnitude that no one causal model could ever encompass it. Even to this day, proponents of such ideas must resort to claims that are nonexistent in nature, such as personifications, selection pressures, adaptive landscapes, niche spaces, arms races, analogies, finch beaks evolution, and outright declaration by fiats, none of which is science.

Teilhard de Chardin was one of the last notable proponents of orthogenesis, the idea that evolution occurs in a directional, goal-driven way.[246] His major philosophical works, *The Divine Milieu* (1957) and *The Phenomenon of Man* (1955), were written in the 1920s and 1930s, but the Jesuits forbade their publication in his lifetime, delaying their publication until the 1950s. Teilhard de Chardin was a "religious person," but his religion differed from what one expects of Christianity, spanning a concept rejected by Darwin and other Darwinists to this day. The Catholic Teilhard de Chardin was not in good standing with the order he belonged to. His faith was not compatible with Catholicism, accounting for the conflict he had with the Catholic Church and the Jesuit order.

Dobzhansky's "Light of Evolution"—No Special Creation

Metaphors that involve light, when used properly, are meant to help understand underlying aphorisms that may be the small underlying truths. Dobzhansky used such a metaphor in his *American Biology Teacher* article. In the context of the article, the metaphor represents two simple ideas: evolution by natural selection provides the understanding for everything that takes place in creation of new life and science, and ideas that are in opposition are to be rejected. Neither is a part of nature or science.

The phrase "light of evolution" represents a worldview and not a process in nature; the way Dobzhansky and the NAS present it, the phrase claims that natural selection integrates all observations about evolution, genetics, biochemistry, neurobiology, physiology, ecology, and other biological disciplines.[247] The NAS and Dobzhansky portray "evolution through natural selection" as the ancient concept of the "quintessence." For many, the quintessence was (and is) God. This concept of quintessence was introduced when people thought that the world was composed only of air, fire, water, and earth. A fifth element, called the "quintessence," was said to bind them all together. The

NAS and Dobzhansky now apparently want to change that religious view to their own to make all observations fit together with their underlying meaning, thereby dictating "how to think" about nature, which is a religious exercise, not one of science. Historically, other metaphors of light also provided a lens for viewing creation and life. One such use of light took place when Jesus used the phrase "light of the world" to describe himself and his disciples in the New Testament:[248]

> I am the light of the world. Whoever follows me will never walk in darkness, but will have the light of life.[249]

For Christians, this aphorism is a quintessence, integrating all observations by one source. Without that quintessence as a worldview, the beliefs of Christians become disjointed from observations and detached from everyday life. Another metaphor of light occurs during Hanukkah, or the "Festival of Lights," which is celebrated for eight days commencing on the twenty-fifth day of the month of Kislev (November–December) to commemorate the victory of the Jews over the Hellenist Syrians in 165 BC.[250] Dobzhansky, as one of the people who helped usher in the modern synthesis, used the metaphor of "light" again. His phrase, "light of evolution," integrates all creation of life when he writes:

> Seen in the light of evolution, biology is, perhaps, intellectually the most satisfying and inspiring science. Without that light it becomes a pile of sundry facts, some of them interesting or curious but making no meaningful picture as a whole.[251]

For Dobzhansky, God is no longer inspiring or satisfying, for the "light of evolution by natural selection" gives meaning to the picture as a whole that is life. He was wrong. He should have known that biology is a subject, not a science. Such a view of evolution as Dobzhansky's does not rise to the occasion of nature, science, biology, or giving meaning to life. The feeling of being satisfied, however appealing to Dobzhansky, is no measure of how creation of new creatures takes place. Dobzhansky simply gave a statement of his religious preference, which for him is apparently is not Christianity. The light of evolution, however portrayed, will always remain poetic, religious, and metaphoric. It will never begin to show cause and effect in nature, which is mandatory for a creation model said to operate in nature's physical world. Regardless of one's feeling, the metaphoric light of evolution exists in the mind, entirely independent of nature's physical operations. That "light" is as applicable to knowing or learning about creation and biology as using evolutionary rose-colored glasses. Dobzhansky tells us about the light of evolution in his biology teachers' article, revealing how it influences his view of creation: it is the sole cause of evolution:

> Evolution as a process that has always gone on in the history of the earth can be doubted only by those who are ignorant of the evidence or are resistant to evidence, owing to *emotional blocks* or to plain *bigotry.*[252]

As is plainly evident from his words, one must either agree with him or be ignorant, have an emotional block, or be a bigot. This is Dobzhansky's science. By implication, it must also be the science of the NAS as it champions him to others. This is Dobzhansky, the NAS's "religious person." He goes on to write:

> By contrast, the mechanisms that bring evolution about certainly need study and clarification. There are no alternatives to evolution as history that can withstand critical examination. Yet we are constantly learning new and important facts about evolutionary mechanisms.[253]

In this quote, Dobzhansky separates evolution from the evolutionary mechanisms (causes) of evolution. Evolution is not a process of creation as Dobzhansky purports it to be in the above quotes. It is an effect—"that which was created." He never names the "mechanisms that bring evolution about"; and thus he acknowledges that evolution is "that which is brought about," as he clearly states. Dobzhansky embraces natural selection in his article but for some reason does not use the correct phrase, "evolution by natural selection," which would correctly show natural selection as the claimed cause and evolution as "that which was created." He is not alone in this misuse of the term "evolution" as a cause. This misuse confuses the term "evolution" in discussions because it may have more than one meaning: first it is an effect; and second, it is said to be a process that causes the effect. In this manner, Dobzhansky makes evolution and natural selection out to be synonymous, which is a significant error.

For Dobzhansky, there is no "evolution by God." There is only "evolution by nature" in the form of natural selection. In practice, Dobzhansky's "light" behaves as an opaque shell that that defines "how to think" about creation of life using a straitjacket of "evolution by natural selection." Dobzhansky's "light" is a metaphor used to demean others if they do not think his way. Other evolutionary mechanisms that are emphasized in recent years include selection pressure, arms races, adaptive landscapes, and need, among others. These mechanisms do not enumerate the parts of nature that may cause creation of specific new genetic material that creates new body parts for new creatures. None of these evolutionary mechanisms exists in nature. Not being causal, they are not responsible for any new creations. (See chapter 10, "Selection Pressure, Arms Races, and Other Incantations" for a detailed discussion.) One incantation, which Darwin endorsed in *Origin of Species*, is a bear changing its body parts gradually and becoming a whale.[254] Michael Denton, author of, *Evolution: A Theory in Crisis,* portrays the impossibility of gradual transitions of a land animal into a whale:

> Let us notice what would be involved in the conversion of a land quadruped into, first a seal-like creature and then into a whale. The land animal would, while on land, have to cease using its hind legs for locomotion and to keep them permanently stretched out backward on either side of the tail and to drag itself about by using its forelegs. During its excursions in the water, it must have retained the hind legs in their rigid position and swim by moving them and the tail from side

to side. As a result of this act of self-denial, we must assume that the hind legs eventually become pinned to the tail by the growth of the membrane. Thus the hind part of the body would have become like that of a seal. Having reached this stage, the creature, in anticipation of a time when it will give birth to its young underwater, gradually develops apparatus by means of which the milk is forced into the mouth of the young one, and meanwhile a cap has to be formed round the nipple into which the snout of the young one fits tightly, the epiglottis and laryngeal cartilage become prolonged downward so as tightly to embrace this tube, in order that the adult will be able to breathe while taking water into the mouth and the young while taking in milk. These changes must be effected completely before the calf can be born underwater. Be it noted that there is no stage intermediate between being born and suckled underwater and being born and suckled in the air. At the same time, various other anatomical changes have to take place, the most important of which is the complete transformation of the tail region. The hind part of the body must have begun to twist on the fore part, and this twisting must have continued until the sideways movement of the tail developed into an up-and-down movement. While this twisting went on the hind limbs and pelvis must have diminished in size, until the latter ceased to exist as external limbs in all, and completely disappeared in most whales.[255]

One may argue that the process of "wolf to whale" transformation is Lamarckian, not Darwinian. In fact, the necessary series of gradual transformations is literally impossible to place into a nature based creation model, which is why such models are never observed. No causal model that addresses nature exists which shows how it takes place. In place of observing the "historical" fossils that show the step-by-step transformations being created by a model, you must imagine them using the same approach of other testimonial models, such as miracles. The model of natural selection does not contain whale parts, organs, biological materials, shapes, processes of nature, or rules of how nature creates, making all the transformations materially inoperative.

The purpose of such claims is to encourage each person to use this gradualist "way to think" about creation: observe fossils and imagine one fossil gradually transforming into the other. It is silently asked that the listener dismiss any competing models, such as special creation. Both natural selection and special creation are testimonial models that require a specific "way to think" about a creature's transitions into becoming a new creature: if you accept one way to think, you deny the other; it is special creation or natural selection, but not both. Both are faith-based. How you think does not involve science, but it does involve religion. Imagining how a four-footed creature is transformed over millennia necessitates one undefined variation being accumulated to one of many undefined accumulations that are forming many undefined organs in a new body. Accepting "that" way to think, you have effectively accepted the religion of an atheist. This is not an endorsement of special creation or atheism, but merely emphasizing what is actually taking place: a choice of belief systems is being made.

Natural Selection Destroys Judeo-Christian Writing—Richard Bozarth (Atheist)

For a theistic evolutionist, in place of God creating new creatures, natural selection is believed to perform all the acts of creation. In a 1978 *American Atheist* article, "The Meaning of Evolution," Richard Bozarth tells us what takes place when one combines a belief in God with natural selection:

> Catholics see pre-Jesus history as "all men born after Adam would be born with the taint of Original Sin, inherited guilt; the race would be blighted and live centuries of longing for a Redeemer." It's none other than Jesus Christ Superstar.

> It becomes clear now that the whole justification of Jesus' life and death is predicated on the existence of Adam and

the forbidden fruit he and Eve ate. Without the original sin, who needs to be redeemed? Without Adam's fall into a life of constant sin terminated by death, what purpose is there to Christianity? None.

Even a high school student knows enough about evolution to know that nowhere in the evolutionary description of our origins does there appear an Adam or an Eve or an Eden or a forbidden fruit. Evolution [by natural selection] means a development (or ascent if one is optimistic) from one form to the next to meet the ever-changing challenges of an ever-changing nature. There is no fall from a previous state of sublime perfection.

Without Adam, without the original sin, Jesus Christ is reduced to a man with a mission on the wrong planet. Death becomes not a divine punishment we require salvation from, but only a natural occurrence as much a part of the normalcy of life as birth. Sin becomes not an ugly fate due to one man's disobedience that we need to be bloodily redeemed of, but only the struggle of instincts learned during millions of years of savagery, trying to adapt to this 10,000 year old infant we call civilization.

Destroy Adam and Eve and the original sin, and in the rubble you will find the sorry remains of the son of god. Take away the meaning of his death. If Jesus was not the redeemer who died for our sins, and this is what evolution means, then Christianity is nothing! And into the void—what? Another religion? I would say yes, for this has been the pattern of history, were it not for what we are building today in the American Atheists. Atheism will be ready to fill the void of Christianity's demise when science and evolution triumph.

Without a doubt, humans and civilization are in sore need
of the intellectual cleanness and mental health of Atheism.[256]

Bozarth's view, like that of others, shows he knows that theistic evolution, which is claiming that natural selection creates new creatures and not God, means the following: There was no Garden of Eden. God did not create the beasts of the earth. God did not create Adam. God did not give dominion over the animals to Adam. Adam did not give names to the beasts. Adam was not created in the image and likeness of God. God did not create Eve from Adam's body. Adam and Eve were not part of the family of God. There was no tree of the knowledge of good and evil. There was no serpent to trick Eve. Eve did not influence Adam into eating the apple from the tree. There was no fall of man. There was no original sin. There is no need for redemption or for Jesus. Jesus did not rise from the dead. There is no Genesis as it becomes myth. The prophets become myth. Communications between humans and God become myth. The Bible becomes a mere story—nice to read, but not important. Atheism becomes the new religion. Morality becomes whatever each individual feels it should be. There is no internal moral control. Force becomes the new moral code. Government becomes the giver of morals and the enforcer.

Dobzhansky's *Light of Evolution*[257] article and the NAS booklet, *A View,* does not mention Bozarth's 1978 article (above) or Blatchford's 1903 article that is referenced below. Neither Dobzhansky nor the NAS mentions Bozarth's classic article about what happens when someone accepts God's use of natural selection to create new creatures, which is to embrace the religion of atheism.

Robert Blatchford (Atheist) Authors: Opposition between Bible and Natural Selection

In 1903, thirty-three years before Richard Bozarth's above-mentioned article in *American Atheist*, Robert Blatchford, an English atheist,[258] also wrote of the neutralization of the Bible when one endorses evolution by natural selection. The effect of theistic evolution, where natural selection (not God) creates new creatures, was well-known in Blatchford's 1903 England, just as it was known in Darwin's time. One may suspect that this is the reason Thomas Huxley privately rejected natural selection but openly defended Darwin. With theistic evolution, God is nowhere to be seen and serves no known role. Blatchford began denouncing organized religion in such works as "God and My Neighbour" in 1903.[259] He writes:

> Let us now turn to the old idea of "The Fall and the Atonement." First, as to Adam and the Fall and inherited sin. Evolution [by natural selection], historical research, and scientific criticism have disposed of Adam. Adam was a myth. But—no Adam, no Fall; no Fall, no Atonement; no Atonement, no Saviour. Accepting Evolution [by natural selection], how can we believe in a Fall? *When* did humans fall? Was it before they ceased to be monkeys, or after?

> Was it when they lived in trees, or later? Was it in the Stone Age, or the Bronze Age, or in the Age of Iron?…And if there never was a Fall, why should there be any Atonement?

> Christians accepting the theory of evolution [by natural selection] have to believe that God allowed the sun to form out of the nebula, and the earth to form from the sun, that He allowed humans to develop slowly from the speck of protoplasm in the sea. That at some period of humanity's gradual evolution from the brute, God found people guilty of some sin, and cursed them. That some thousands of years later God sent His only Son down upon the earth to save humanity from Hell. But evolution shows humans to be, even now, an imperfect creature, an unfinished work, a building still undergoing alterations, an animal still evolving.

> Whereas the doctrines of "the Fall" and the Atonement assume that he was from the first a finished creature, and

responsible to God for his actions. This old doctrine of the Fall, and the Curse, and the Atonement is against reason as well as against science.

The universe is boundless...Are we to believe that the God who created all this boundless universe got so angry with the children of the apes that He condemned them all to Hell for two score centuries, and then could only appease His rage by sending His own Son to be nailed upon a cross? Do you believe that? Can you believe it?...if the theory of evolution [by natural selection] be true, there was nothing to atone for, and nobody to atone. *Man has never sinned against God*...There was no creation. There was no Fall. There was no Atonement. There was no Adam, and no Eve, and no Eden, and no Devil, and no Hell.[260]

The NAS tells us those theistic evolution beliefs are "not in disagreement with *scientific explanations* of *evolution*,"[261] but they never explain why, perhaps because theistic evolution is an atheistic religion. Theistic evolution beliefs, as shown by Bozarth and Blatchford, disagree with Judeo-Christian beliefs, beliefs in Genesis, acceptance of the Bible, and miracles. The NAS does not elaborate upon or in any way explain their position in their booklet *A View*. The "religious person" they endorsed belonged to the atheist religion. They should have discussed Bozarth and Blatchford in their booklet *A View*, but they did not even try.

Thomas Wollaston and Creating Only Once

While never observed in nature, natural selection is said to be a "continuous" series of creations, which Eldredge and Gould titled "phyletic gradualism." In contrast, God created natural selection only once, according to Darwin, just as the models of Newton's gravity and Archimedes' lever were created only once. Another "only once" creation, according to Darwin, is when the *first life forms* were created by God, just as his grandfather had declared in *Zoonomia*. Natural selection created all life afterward. Thomas Wollaston criticized this view of "God creating only once" when he reviewed the *Origin of Species* shortly after Darwin published it. Giving his views about Darwin's claim of God creating only once and then stopping all acts of creation, Wollaston writes:

> To our mind, the wonder consists in the act [of creation] at all, and not in the number of times that it may have been repeated: for a Being that can create may surely do so as often as he pleases; and we have no right therefore to limit that act,—at any rate on the question of its probability; for, if we admit that it has been exerted so much as once, there is no *a priori* reason why it should not have been a million times repeated, or why, if he had so willed it, it might not, at some period or other, have been in constant operation.[262]

The question that comes to mind is simple: by what right or theory does anyone limit the faith in God and his actions at all if he or she endorses God's creating even once? Science doesn't back such claims; only competing faith systems do. Darwin's faith limited his view of God to one creation of the first life forms. Others' views of creation, including those creations in Genesis, do not limit the number of creations God could make or how often He could create new creatures. No such right or theory exists in nature by which to limit how often God creates anything. Only an opposing faith system and opposing worldview claims such a right. This faith system holds, as its core tenet, complete opposition to special creation, but there is no science in that view or belief. It could be asked: by what right does anyone have to limit God to no creations? No matter what the answer, it is not science. It is a religious answer. As with the rest of his works, there is no science in the answer.

This is also true for the NAS creation model of theistic evolution, where, under the guise of what they consider to be science, they claim that God creates natural selection and does not create any creatures, with the possible exception of the first ones: this belief holds that God created only life only once. No knowledge exists in nature or science by which they make the claim. In Darwin's case, the knowledge he used to make his claims did not exist in science but could be found in opposing faith. Even if it were claimed that God never created any life and spontaneous generation was responsible, by what right would such a claim be made and used to declare that God did not create at all? Darwin and the NAS limiting of the biblical God, or any God, to their creation view is an arbitrary invention of their own religion. It is by faith that one claims God created natural selection. It is by blind faith that one limits God's actions of creation to only the act of natural selection and the first life forms. One cannot possibly determine that God created only one life form or one thing (natural selection) except by initiating a new faith system, which Darwin did—and the NAS continues to do. That "create one time" God is arbitrary and is in conflict with Genesis. But one would not know that from the NAS booklet, *A View.*[263] The belief that God created only "one-time", whether it is the first life forms or natural selection, is the God endorsed in theistic evolution. There is no science to this belief.

No "scientific explanations of evolution"[264] may contain any injection of God, either for or against His involvement. Newton knew that. Mendel knew that. Pasteur did as well. Those who make such claims about evolution, to include God in some way, must explain how that constitutes science. If it does not constitute science, they why discuss God at all? With theistic evolution, either God creates or natural selection creates, but not both: not in combination, not in sequence, and not in parallel. The choice of God (any type of God) or natural selection marks a fork in a path where one excludes the other. Natural selection is entirely atheistic. Just as importantly, each one excludes nature and science. Both are based on faith. When God is combined with natural selection, He becomes unnecessary and inactive; God is merely an add-on, a likely deception to those believing in Genesis, a belief necessitating faith that is no different in kind than the faith in Genesis.

This "God only created natural selection" view is, at best, as close to the religion of atheism as one may get without openly endorsing it. In practice, it is the religion of atheism. The NAS does not explain or mention God's role[265] in creation or how faith fits into the NAS's view of a "scientific explanation of evolution."[266] That explanation would prove interesting, but is not given. Perhaps this is because no explanation of the cause of evolution involves science. Every explanation of the cause of evolution is faith-based, as demonstrated by the constant need for testimonial attributions. The only question involving creation and evolution is this: which faith is used, Darwinism or some competing one?

Animals Yes, People No

The question of how a person declares another person's religion to be wrong is not science. Some of Darwin's friends, such as Wallace and Lyell, thought that the creation of human beings by natural selection could not be dealt with in the same way as animal species; Lyell never believed in natural selection and Wallace only partly. Lyell's view rejected Darwin's gradualism.[267] It also rejected Genesis, which shows that God created both people and animals as shown in Genesis chapter 2:

> [7]And the LORD God formed man *of* the dust of the ground, and breathed into his nostrils the breath of life; and man became a living soul.[268]

> [19] And out of the ground the LORD God formed every beast of the field [special creation], and every fowl of the air; and brought *them* unto Adam to see what he would call them: and whatsoever Adam called every living creature, that *was* the name thereof.

> [21] And the LORD God caused a deep sleep to fall upon Adam, and he slept: and he took one of his ribs, and closed up the flesh instead thereof;

> [22] And the rib, which the LORD God had taken from man, made he a woman, and brought her unto the man.

[23] And Adam said, This *is* now bone of my bones, and flesh of my flesh: she shall be called Woman, because she was taken out of Man.[269]

Wallace's and Lyell's separation of creation into one part for natural selection and one part for God is denying the Bible. The only possible means of declaring such a separation of creations is by using two faith systems: one for natural selection creating animals and a second (God) for the creation of humans. In this way, a person's worldview and religion establishes, for him or her, a new religion that rejects Genesis, the Bible, and the God of Genesis. This is a hybrid creation model composed of theistic evolution and special creation by a non-Genesis God.

Summary of Theistic Evolution

Theistic evolution is atheism in practice. To theistic evolutionists, God is not present in the world or involved with creation of new creatures. The claim that God created natural selection is not based in science but in the religious tenets of *Darwinism.* Theistic evolution is unrelated to science, Genesis, and the Bible, including the New Testament. The claim that God used natural selection to create new creatures is in contradiction to both the Bible and science. It merely means endorsing a religion whose greatest tenet is "No special creation allowed." That belief has consequences: there is no Garden of Eden, no Adam, no Eve, no forbidden tree of knowledge, no apple, and no serpent. *Genesis* becomes myth. The NAS does not write about these consequences. Instead, it merely claims that a religious person can hold that God created the universe and the various processes driving physical and biological evolution. According to Hodge, one of the greatest exponents and defenders of historical Calvinism in America during the nineteenth century:

The conclusion of the whole matter is that the denial of design in nature is virtually the denial of God. Mr. Darwin's theory does deny all design in nature, therefore, his theory is virtually atheistical; his theory, not he himself. He believes in a Creator. But when that Creator, millions on millions of ages ago, did something,—called matter and a living germ into existence,—and then abandoned the universe to itself to be controlled by chance and necessity, without any purpose on his part as to the result, or any intervention or guidance, then He is virtually consigned, so far as we are concerned, to non-existence. It has already been said that the most extreme of Mr. Darwin's admirers adopt and laud his theory, for the special reason that it banishes God from the world; that it enables them to account for design without referring it to the purpose or agency of God.[270]

As Hodge writes, believing in natural selection is atheism. Some Darwinists consider themselves religious, as is the case with theistic creationists, but the religion has nothing to do with science, nature, or the Bible. Believing in natural selection is, in practice, atheism. Hodge tells us about Darwin's God—about His removal from creation, His having no communications with man, and his not caring about what takes place on earth. Hodge continues giving the consequences of believing in natural selection and believing in a God:

There are men who are constrained to admit the being of God, who depart from the Scriptural doctrine as to his relation to the world. According to some, God created matter and endowed it with certain properties, and then left it to itself to work out, without any interference or control on His part, all possible results. According to others, He created not only matter, but life, or living germs, one or more, from which without any divine intervention all living organisms have been developed. Others, again, refer not only matter and life, but mind also to the act of the Creator; but with creation his agency ceases. He has no more to do with the world than a ship-builder has with the ship he has constructed, when it is launched and far off upon the ocean. According to all these views a creator is a mere *Deus ex machina*, an assumption to account for the origin of the universe.[271]

Hodge has described the God of the theistic evolutionist, an injection of God to move the story of creation forward when the writer sees no other way out. This God may incorrectly be held as the deist God, but the characteristics of Darwin's God are different. By defining "chance" in the first part of natural selection so that not even God knows what is to take place, Darwin made God less than all-knowing and less than all-powerful. He certainly is not the God of Genesis or a deist God.

If one exercises his or her faith in this manner, it is no different *in kind* than other kinds of faith, with the added burden of natural selection being self-contradictory: it has no physical components in the model to operate in nature and cannot be science. Thus, a believer in God who uses natural selection must have faith in such a God and faith that a model without physical representation in nature must, somehow, have a reality when everything contradicts that belief.

Natural selection has no parts of nature, just as miracles have no parts of nature. The two models operate in the same way, by testimony. This makes it impossible for either model— miracles or natural selection—to create any creatures or body parts using nature's physical world and physical processes. One knows no more about nature or creation after using natural selection than believers in Genesis know after using special creation. The feeling by theistic evolution supporters that natural selection is in need of some sort of association with God is without merit in science or nature. The NAS religious person who uses God and natural selection has forfeited both science and biblical Genesis in the process. Such people can be "religious," like Darwin, Dobzhansky, or the people of the NAS, but that does not represent nature, biology, or science. It represents religious faith. This raises that question: why is natural selection being taught in classrooms as biology or science when it is neither? Why isn't it being taught solely in a classroom for religious studies?

CHAPTER 5

DARWINISM DEFINED

The term "Darwinism" first applied to Erasmus Darwin (1731–1802), who wrote about naturalistic evolution long before his grandson, Charles. He preceded Lamarck in working on creation and evolution and was the first Briton to explicitly write about evolution by naturalistic causes. His main prose on naturalistic evolution appears in the 1796 volume of *Zoonomia*, in which he wrote about the descent of life from a common ancestor; sexual selection; the analogy of artificial selection as a means to understanding descent with modification (animal breeder analogy); a basic concept of what we now refer to as homology; and the idea that God created the first life-forms but had no further role in creation, which made him a theistic evolutionist with the eternally absent deist God. His thoughts on the diversity of life and evolution also appeared in the *Loves of the Plants* (first published in two parts in 1789 and 1791) and in his last work, *The Temple of Nature,* published posthumously in 1803.[272] The meaning of Darwinism, although it has changed over time, has its core remaining essentially the same: no special creation allowed. Samuel Butler in the 1911 edition of his book, *Evolution, Old and New,* writes about the early use of the term Darwinism:

> Then follows a passage [in Paley's book *Natural Theology*] which is interesting, as being the earliest attempt I know of to bring forward an argument against evolution, which was, even in Paley's day [1743–1805], called "Darwinism," after Dr. Erasmus Darwin its propounder. The argument, I

mean, which is drawn from the difficulty of accounting for the incipiency of complex structures. This has been used with greater force by the Rev. J. J. Murphy, Professor Mivart, and others, against that (as I believe) erroneous view of evolution which is now generally received as Darwinism.[273]

Even in its early usage, Darwinism referred to creation without God, without special creation, and without miracles. This would, of course, mean creation takes place by nature's components and processes or "naturalism." In another example, Butler writes about the use of the term Darwinism:

"What!" says Colerirge, in a note on Stillingfleet, to which Mr. Garnett, of the British Museum, has kindly called to my attention, "Did Sir Walter Raleigh believe that a male and female ounce [i.e., panther] (and if so why not two tigers and lions, etc.?) would have produced in course of generations a cat, or a cat and a lion? [this is the common ancestor or creatures changing into new forms of creatures over time]. This is Darwinising with a vengeance."[274]

Ernst Mayr writes that T. H. Huxley coined the term "Darwinism" in 1864,[275] but the Samuel Butler account of the term precedes that date by about sixty-two years. Alfred Russel Wallace, the "cofounder of natural selection," published a whole volume titled *Darwinism* in 1889, which he used as "an exposition of the theory of natural selection with some of its applications." It was meant to be complimentary to Darwin. The term has been used in different ways by different users to this day. Adrian Desmond, in his 1994 book, *Huxley: From Devil's Disciple to Evolution's High Priest*, explains the formation of Darwinism and those who fell under the meaning of the term:[276]

The word evolution became common currency in the 1870s. For a decade, "Darwinism" had been the term to confer a multitude of sins. But the Darwinians had been an undifferentiated lot, who clustered for strength and tended to make little distinction between life's genetic development and Darwin's explanation of natural selection.[277]

Even then, the term "Darwinism" was not favorably used, as shown in the following passage:

> Attracting radicals, atheists, socialists, and free-traders, "Darwinism became notorious as much for the friends it kept as for its political enemies." But at the end of the 1860s the groups were peeling off, emphasizing their internal differences. It was no coincidence that as the word "evolution" came in and the dissidents fell out, Huxley devised his "agnosticism" to legitimate only the secular and naturalistic end of the spectrum.[278]

There were many Darwinists that accepted "common descent" but rejected natural selection. Leading Darwinists like Huxley and Lyell never believed in natural selection.[279] Desmond writes, "Huxley diverged on key points."[280]

> Where Darwin had nature select from tiny variations, Huxley had been happy with larger jumps; where Darwin transported his species to islands by wind and raft, Huxley had invoked drowned continents. At times, all they seemed to share was a faith in evolutionary naturalism [godless—no special creation].[281]

The meaning of Darwinism, like the meaning of natural selection, depends on the user and his or her views.[282] Neither Huxley nor, presumably, Lyell ever endorsed Darwin's complete gradualism.[283] In his article, "Lyell's Views on Organic Progression, Evolution and Extinction," A. Hallam shows that Lyell, in fact, never did endorse natural selection, neither earlier nor later in his life.[284] Neither Wallace nor Lyell thought that human beings could be dealt with in the same way as animal species.[285]

Wallace, the co-discoverer of natural selection, thought that natural selection could create much of the body of humans but could not create humans' "higher qualities," which required the intervention of God, though he never specified what type of God. Wallace's tests showed this was true (see chapter 11, "Testing Miracles and Natural Selection"). Wallace did explain why he attributed some creation of humans to God and some to natural selection; his explanation appeared in his 1869 article, "The Limits of Natural Selection as Applied to Man."[286] Just what type of God Wallace believed in is not clear. Perhaps it was pantheism, deism, an intelligent agent, or a life force. But one thing is certain: Wallace did *not* have the Christian biblical God in mind.

Core Belief of a Darwinist

For some Darwinists, God may have been a creator of creation models such as natural selection. For these people, once the model (such as natural selection) was created, the model created all the creatures in existence, but God did not. Charles Darwin was this type of Darwinist. Darwin's American friend, Asa Gray, was another type of Darwinist. For Gray, God directed variations in natural selection; but this was still not the biblical God who created life by special creation and talked to Adam and Eve or sent His only son Jesus to remove original sin. For some, like Wallace, natural selection could not create people's higher qualities, but it did create the rest of the human body and the bodies of all other creatures. For many, natural selection was synonymous with atheism. Many types of people, even those holding conflicting beliefs, were called Darwinists. Ernst Mayr writes about the one core worldview needed to be a Darwinist.

> There is indeed one belief that all true original Darwinians held in common, and that was their rejection of creationism, their rejection of special creation. This was the flag around which they all assembled and under which they marched... Nothing was more essential for them than to decide whether evolution is a natural phenomenon or something controlled by God.[287]

Mayr claims it is either "evolution by natural phenomenon" or "evolution by God" as the creation model. He continues:

> The conviction that the diversity of the natural world was the result of natural processes and not the work of God was the idea that brought all the so-called Darwinians together in spite of their disagreements on the other Darwin's theories, and in spite of the retention of some of them…of other theological arguments. This situation was quite well understood in the post-*Origin* period and that is why at that time, for Darwin's opponents, Darwinism simply meant denying special creation and replacing it with the theory of evolution [by natural selection] and in particular the theory of common descent [through natural selection].[288]

Without the one core idea in its meaning, Darwinism and natural selection would have become minor footnotes in history. Darwinism was kept alive by that core meaning, then as now. It is the social glue of all Darwinists, despite their other qualities and disagreements that would otherwise drive them apart. You cannot observe this core view in the model of natural selection, revealing again that the model is inoperative, taking on only the characteristics of those using it, for it lacks any of the characteristics of science, such as describing the operative physical parts of nature, the consistent repeatability of the model in nature, and the absence of attribution and testimony. Natural selection's terms allow the core belief of a Darwinist to be attributed to it, thereby attributing nature to the model and thus making it appear supportive of the Darwinist's own worldview. All models whose terms are independent of nature possess this attributive characteristic, but no models of science operate in this manner.

The X-Club Agenda and Darwin

To advance some commonly held causes, Huxley decided to start a private "dining club" of his own that would bring together a significant bunch of like-minded activists. After failing to agree on a name for the group, they chose the "X-Club," which gave the advantage of committing to nothing, as Spencer said approvingly. The only rule of the club was that there were to be no rules. The obvious aims of the X-Club included personal friendship, devotion to science,[289] and promotion of natural selection against special creation, along with the aim of being "pure and free, untrammeled by religious dogmas."[290] Over dinner, the members would catch up on gossip and scheme about the things they considered to be scientific. Soon they became an important, informal pressure group and focused their attentions on the "scientific" administration. They began promoting the liberal naturalism associated with Darwin's theory.[291] Although any one of them might not agree with natural selection, they agreed with its meaning; they agreed that it provided the means by which they would repulse special creation. The X-Club was in no small part responsible for Darwin's becoming noted, and the members actively promoted Darwin and natural selection.

Huxley's first dinner party included friends like Hooker, Spencer, Tyndall, Lubbock, and Busk, as well as two new acquaintances, the mathematician Thomas Archer Hirst and Edward Frankland, a clever chemist at the Royal Institution. For the next party, Huxley added William Spottiswoode, another mathematician; with Huxley, they numbered nine in all.[292] One by one, the members infiltrated every government panel and committee that dealt with scientific affairs,[293] where they made a lasting impact. In their wake, Darwin and natural selection would begin appearing in textbooks as science. Natural selection served as the talking point for this club, allowing its members to speak their views about creation and evolution.

In return for Darwin's assistance and cooperation, the members in the X-Club counted on using Darwin's influence, and they tirelessly asked for it;[294] with it, the influence and agenda of the X-Club progressed. They demonstrated their beliefs and agenda, as well as that of Darwin, through their defense of a minister by the name of John William Colenso. Colenso's liberal reevaluation of the first five books of the Old Testament proved more than most parsons could take. Colenso regarded the earliest sections of the Bible as little more than a collection of ancient historical documents that had accumulated errors and misreadings over the centuries. He was accused of heresy.[295]

A petition circulated in progressive philosophical circles, "a declaration in favor of freedom of opinion and defending the rights of Bishop Colenso," as Darwin's son, Erasmus, called it. Erasmus and Darwin both signed it,[296] effectively endorsing the destruction of Genesis—hardly an undertaking of science. By so signing, Darwin showed his denial of the creation, Genesis, and God. Darwin also showed his support of atheism.

Huxley's band of "X Clubbers" invited the "Bible-killing" (so named) Colenso to dine, and in 1864 they set up a subscription fund to provide a sum of money should he be defrocked, which some of them half hoped might happen. In 1865, Colenso flatly contradicted the story of creation and the deluge on "scientific" grounds. It was clear that X-Clubbers, and Darwin, with his signing of the petition to defend Colenso, had an agenda unrelated to biology, science, or nature. The agenda for the X-Club formed the agenda for Darwinists to this day: to use natural selection as the testimonial vehicle to defeat all competing models of creation. Charles Hodge, principal of Princeton Theological Seminary, highlighted the motivating force for Huxley:

> Mr. Huxley calls believers in the Scriptures, and (apparently) believers in a personal God, bigots, old ladies of both sexes, and bibliolators, fools, etc., etc., etc.[297]

In recent times, Ernst Mayr describes what Darwinism meant in Darwin's time:

> Immediately after 1859, Darwinism meant a rejection of special creation. If someone rejected special creation and adopted, instead, the inconstancy of species, common descent, and the incorporation of man into the general evolutionary stream, he was a Darwinian.[298]

There were great differences between Darwin and the other "Darwinists," such as Huxley, Lyell, Wallace, and Gray, on other aspects of evolutionary theory. But these differences were of minor importance in the 1860s, because the foremost meaning of Darwinism at that time was the *rejection of special creation*, together with the adoption of inconstancy of species, the theory of common descent, and (excepting Wallace) that incorporation of man into the animal kingdom[299]...Neither natural selection nor any special theory of speciation, nor even one's belief in gradual versus saltational evolution, had any relevance to whether at that time one was considered a Darwinian or not. [Italics added][300]

That core meaning was evident with Erasmus Darwin. For him, creation was by naturalistic means was important—it was his religious belief, not a scientific one. There are those who accept natural selection for the creation of animals, as Wallace did, but they exclude the creation of humans' "finer characteristics" from the actions of natural selection. This mixture of God creating "some of man" and natural selection's creating other parts of man poses problems, as Hodge tells us:

> The minds of some men...are so constituted that they can pass from the theory that God does nothing, to the doctrine that He does everything, without seeing the difference.[301]

As Hodge shows, God's role is determined by ones views, their beliefs, and their worldviews. It is by a person's religious beliefs that such determinations are made, not by nature or science. Hodge continues:

> Mr. Russel Wallace, the companion and peer of Mr. Darwin, devotes a large part of his book on "Natural Selection," to prove that the organs of plants and animals are formed by blind physical causes. Toward the close of the volume he teaches that there are no such causes. He asks the question, What is Matter? And answers, nothing. We know, he says,

nothing but force; and as the only force of which we have any immediate knowledge is mind-force, the inference is 'that the whole universe is not merely dependent on, but actually *is*, the will of higher intelligences, of one Supreme Intelligence. This is a transition from virtual materialism to idealistic pantheism. The effect of this admission on the part of Mr. Wallace on the theory of natural selection is what an explosion of its boiler would be to a steamer in midocean, which should blow out its deck, sides, and bottom. Nothing would remain above water.[302]

Wallace always included natural selection in the creation of animals, but with some parts of humans, he changed horses midstream, including God and excluding natural selection. With Wallace, as with Darwin, natural selection took on *his* worldview and *his* beliefs that God acted in one creation role and natural selection in the other creation role. Natural selection does not indicate any preference one way or the other, illustrating that the model and attributions made to it assume the user's worldview, not the roles of nature—a characteristic not found in the models of science. Creation by attribution, not creation by nature or God, is the only way natural selection functions.

Summary of "What Is Darwinism?"

To be a Darwinist, one must reject special creation. One must believe that creation does not take place through God's intervention using special creation, which is sometimes called "independent creation" or "separate creation." Darwinism today has much in common with the 1800s Darwinism of Erasmus Darwin as well as the Darwinism of 1859, when *Origin of Species* was published. The primary goal of Darwin and Darwinists then, as is the primary goal for Darwinists today, was to replace every other model of creation in competition with it, notably special creation and miracles as in Genesis. Evolution is the many fossils appearing in the ground, gathered and organized as a table of evolution, which shows different perspectives of creation. The fossil creatures are not what fuels the public debate about evolution; the debate is fueled by the cause of new creatures' creation, which, for Darwinists, is natural selection, despite natural selection's independence of nature.

Janet Browne tells us why Darwinism is used to represent evolution:

> During the 1870s, Darwin became the most famous naturalist in the country…The term "Darwinism" almost by default, covered all kinds of evolutionism and unfairly eclipsed the work of Huxley, Wallace, and others.[303]

There is a reason for this association of Darwin and Darwinism with evolution. His name was rapidly becoming part of the culture and along with his name, almost synonymously with it, "evolution by natural selection" was becoming part of the "richly varied world of nineteenth-century culture."[304] Browne continues:

> Even his own book [*Origin of Species*] was now apparently more discussed than read.[305]

It is easy for information to be incorrectly or incompletely gleaned as secondhand information under the name of Darwin with people's incomplete views of natural selection and evolution. Browne notes:

> It has been so easy to learn something of the Darwinian theory at second-hand, that few have cared to study it as expounded by its author, Wallace said accurately enough after Darwin's death.[306]

Darwin's name had become a symbol and "Darwin's theory," while not be technically understood had become a metaphor that symbolized evolution. Browne notes Darwin's reaction to his "star" status:

Slowly, and somewhat uncomfortably, Darwin recognized that his celebrity, however much he disliked it, worked to his theory's advantage.[307]

The Darwin name has become an icon, starting with Erasmus Darwin. It has become an industry to promote one view of evolution – the "cause" of it. The industry thrives because the name "Darwin" is easy to pronounce and easy to associate with the phrases "creatures change slowly," "survival of the fittest," and "no special creation allowed." But what is easiest is also very unscientific, for it fails to portray vital information about Darwin's theory—namely, that it does not operate in nature, operating instead in the imagination. Models without the ability to show what operates in nature and how, must function by attribution and testimony, which is how natural selection functions. Darwin's model of nature does not show creation of one new creature or its new body parts taking place, slowly or rapidly, in multiple steps or in one step: all that creation activity must be inferred and believed, essentially forming a belief system.

Charles Darwin's supporters used both natural selection and Darwin himself as the easy handles to represent beliefs in naturalistic creation and opposition to special creation. The claims that God is not involved with creation by natural selection may as well be synonymous for "no special creation" or atheism. Atheism is not the point here, but natural selection's claim as science is the point: it is a misleading claim. In the United States of America, freedom of religion is recognized as the law of the land, but publishing a religion renamed as science, as in the case of natural selection, evades that First Amendment freedom and undermines it. Students do not get the freedom to choose their religion, for they are spoon-fed atheism in the biology and science textbooks under the name of natural selection, which has become the new religion of the United States. There is no science involved.

CHAPTER 6

STRANGE BEDFELLOWS

Introduction: The Meaning of It All

The phrase "survival of the fittest" is not original with Darwin, who used that phrase for the first time in the last two editions of *Origin of Species*. The phrase, as Darwin used it, was originated by Herbert Spencer and originally used in a political context rather than a biological one. Despite it commonly being attributed to Spencer's 1851 book *Social Statics, or The Conditions Essential to Happiness Specified, and the First of Them Developed*, he did not coin the phrase until his 1864 *Principles of Biology*. He later applied "survival of the fittest" to economics and, with Darwin, incorrectly to biology. This phrase does not have the remotest connection with science; rather, it is used as a key tenet of the so-called social Darwinism, which is, in fact, natural selection or survival of the fittest applied to people, but otherwise no different than that applied to other creatures, such as plants, insects, fish, and animals. The strength and weakness of the phrase is that it can be applied to a diverse number of situations and sound meaningful. In reality, the phrase is a metaphor, not having any existence in nature or science. If it sounds and reads as if it were very real, that would be an error. Spencer writes:

> But this survival of the fittest implies multiplication of the fittest. Out of the fittest thus multiplied, there will, as before, be an overthrowing of the moving equilibrium wherever it presents the least opposing force to the new incident force. And by the continual destruction of the individuals that are

the least capable of maintaining their equilibria in presence of this new incident force, there must eventually be arrived at an altered type completely in equilibrium with the altered conditions. This *survival of the fittest*, which I have here sought to express in mechanical terms, is that which Mr. Darwin has called "natural selection, or the preservation of favoured races in the struggle for life."[308]

That organisms which live thereby prove themselves fit to live, in so far as they have been tried; while organisms which die, thereby prove themselves in some respects unfitted for living.[309]

It follows from survival of the fittest that any creature alive is fit. The term "fitness" is not tied to nature. It has nothing to do with nature. It is not dependent upon nature, as are other terms found in nature such as temperature, distance (feet, meters), weights (pounds, kilograms), brightness (lumens), volume (cubit feet or meters), time (minutes, years), germs (rabies, anthrax, cholera, smallpox), or Mendel's genetic traits (color, roughness). Fitness is attributed to nature by the user, much like the other terms that also do not exist in the physical world, such as "far," "near," "hot," and "cold." The person using the terms is the link between fitness and creatures. This fitness has a number of definitions, such as the number of offspring, personal qualities (speed, strength, size, agility), and abilities such as flying, swimming, burrowing, and intelligence. Arguing with someone's definition of a term that is independent of nature, such as fitness, is no different from arguing about whether something is cold or warm. Either claim can be a correct way to view nature for any person as a personal view. But neither view involves nature or science.

One may agree or disagree with Darwin about the meaning of fitness when he defined it in *Origin of Species* as "let the strongest live and the weakest die,"[310] with his view of fitness meaning "the strong" and the unfit meaning "the weak." The survival of the fittest is the entire theme of evolution by natural selection,[311] even when applied to people. The model of natural selection offers the morality of "following one's feelings" as a moral basis for whatever actions a person chooses. When the phrase is used by a government, the people in control of that government define the meaning of "fit" and "fitness." Arguing about that definition becomes an arguing with the government.

Spencer believed that competition was the law of life and resulted in the survival of the fittest.[312] "Society advances," Spencer writes, "where its fittest members are allowed to assert their fitness with the least hindrance." He goes on to argue that the unfit should "not be prevented from dying out." Many interpretations are possible from this concept. Eugenics was born from it.

Making People Fitter: In the United States

Nazi Germany practiced forced sterilization of "undesirables," consistent with their totalitarian practices and their selection of the fittest. However, many find it surprising that forced sterilization was first practiced in the United States. In both cases, those deemed "undesirable" could be forcibly sterilized. If survival of the fittest is science, then sterilization practices were thereby provided a *scientific* license: who could deny science? Neither principles in evolution by natural selection[313] nor rules derived from survival of the fittest conflict with the actions of *any* user of the model: the rules and actions are what they say they are. Despite the term "fittest" having no definition that is dependent on nature, many think that arbitrarily assigning definitions to "fitness" on a case-by-case basis is a satisfactory practice. The difficulty with such a practice is that anyone can be defined as unfit, even for their philosophy or religion. "Fitness" is an arbitrary term. In the hands of a government, it can be lethal. The term "eugenics" was coined in 1883 by Francis Galton, the cousin of Charles Darwin.[314] Galton believed in eugenics. He used "survival of the fittest" as the basis of his Eugenics Society.

> The United States was the first country to concertedly undertake compulsory sterilization programs for the purpose of eugenics. The heads of the program were avid believers in eugenics and frequently argued for their program. In the end,

over 65,000 individuals were sterilized in 33 states under state compulsory sterilization programs in the United States.[315]

Galton first published his eugenic ideas in 1865—well before he coined the word itself—in a two-part article for *Macmillan's Magazine*, which he subsequently expanded into a book, *Hereditary Genius*, published in 1869. In addition to being Darwin's cousin, Francis Galton was the son of a wealthy, influential family. In 1869, in *Heredity Genius*, he followed the lives of several accomplished men from what he considered good families. These good families, Galton claimed, would more likely produce intelligent, talented offspring.[316] It was well-known that by careful selection, farmers and flower fanciers could obtain permanent breeds of plants and animals strong in particular characters. Galton wondered, "Could not the race of man be similarly improved? Could not the undesirables be got rid of and the desirables multiplied?[317] Galton had expected eugenics to provide a secular substitute for traditional religion.[318]

Charles Darwin thought so highly of his cousin Galton that he practically canonized him in *The Descent of Man*, repeatedly glorifying him and his work[319]. In the 1910 United States, Mrs. Harriman, "a socialist activist with a liberal bent,"[320] using her late husband's immense railroad fortune, funded the establishment of a Eugenics Record Office on seventy-five acres of land, which she bought in Cold Spring Harbor, New York, on Long Island's north shore. Supporters of eugenics included notables such as Alexander Graham Bell, Luther Burbank, and Andrew Carnegie. In 1907, the first sterilization law had passed in Indiana.[321] By 1927, a forced sterilization test case, Buck v. Bell, had reached the United States Supreme Court where, by a vote of eight to one, forced sterilization was upheld. Justice Oliver Wendell Holmes wrote the majority opinion. By the end of the 1920s, sterilization laws were on the books of twenty-four states. Though now severely restricted by federal regulation, these laws are still on the books of twenty-two states today.[322]

The Nazis used survival of the fittest to eliminate procreation by individuals whom they considered "unfit" to live in society, either physically, mentally, socially, or by religion. The theories of Darwin and Galton that were used by the Nazis, fascists, and communists are matters of personal choosing, not of physical fact operating in causal models of nature. What one person finds objectionable may seem necessary to another, and the words in the model offer no means of differentiation. This demonstrates the very nature of testimonial models in general, and survival of the fittest or natural selection in particular. The Supreme Court sanctioned the United States being the first country to impose forced sterilization on its citizens, despite the legitimate claims that the terms used to determine who should be sterilized were arbitrary (i.e., independent of nature). Building laws upon terms that are independent of nature amounts to criminals being arbitrarily determined.

Making People Fitter: In Germany

Spencer used survival of the fittest to argue for limited government, free markets of individuals, and capitalism. Hitler used survival of the fittest to argue for its complete opposite: socialism. Again, the model of survival of the fittest offers no means of determining what is right or wrong. Hitler's form of socialism encompassed Nazi totalitarianism, differing very little from communist socialists' totalitarianism. Hitler endorses the phrase "survival of the fittest" in *Mein Kampf, which means* "my struggle." He writes:

> It is in this activity on the part of the membership body, guaranteed by the process of natural selection, that we are to seek the prerequisite conditions for the continuation of an active and spirited propaganda, and also the victorious struggle for the success of the idea on which the movement is based.[323]

> [W]e give free play to the natural process of selection which brings forward the ablest and most capable and most industrious.[324]

The terms used above by Hitler—"ablest," "most capable," and "most industrious"—are definitions commonly attributed to the term "fittest." These terms are independent of nature and well aligned to Darwin's model of survival of the fittest. Personification is also used in the quote, just as Darwin had used it. This open-ended, theoretical license granted by its independence from nature underpins Hitler's actions. When reading Hitler's theory in *Mein Kampf*, a discussion of which follows, there is no doubt that Darwin's ideas are being discussed.

Hitler: *Mein Kampf* and Darwin

Because of the broad interpretation base of many terms' meanings in evolutionary biology, each person may personally adopt his or her own views. Hence, many prominent people supported eugenics without close examination of its meanings, accepting it with apparent innocence of what it meant and could mean. At its peak of popularity, a wide variety of prominent people supported eugenics, including Winston Churchill, Margaret Sanger, Marie Stopes, H. G. Wells, Theodore Roosevelt, George Bernard Shaw, John Maynard Keynes, John Harvey Kellogg, Linus Pauling, and Sidney Webb.[325] The First International Congress of Eugenics in 1912 was supported by many prominent persons, including its president, Leonard Darwin (the son of Charles Darwin); honorary vice-president Winston Churchill, then First Lord of the Admiralty and future prime minister of the United Kingdom; Auguste Forel (famous Swiss pathologist); Alexander Graham Bell (the inventor of the telephone); and Darwin's cousin, Francis Galton, among other prominent people. In the following quotes from *Mein Kampf*, Hitler also endorses eugenics, the forced sterilization of "unfit" or "undesirable" people. Hitler writes:

> Every crossing between two breeds which are not quite equal results in a product which holds an intermediate place between the levels of the two parents. This means that the offspring will indeed be *superior* to the parent which stands in the biologically lower order of being, but not as high as the higher parent. For this reason it must eventually succumb in any struggle against the higher species.

Such mating contradicts the will of Nature toward the *selective* improvements of life in general. The favourable preliminary to this improvement is not to mate individuals of higher and lower orders of being but rather to allow the complete triumph of the higher order [eugenics]. The stronger must dominate and not mate with the weaker, which would signify the sacrifice of its own higher nature.

Only the born weakling can look upon this principle as cruel, and if he does so it is merely because he is of a feebler nature and narrower mind; for if such a law did not direct the process of evolution then the higher development of organic life would not be conceivable at all.[326]

The terms used by Hitler—"stronger," "weaker," "superior," "nature," "weakling," "favourable," "feebler nature," "narrower mind," "the higher order," "biologically lower order of being," "higher nature"—are terms commonly used in evolutionary biology. They are not, however, found in nature or science as they are independent of nature, just as the terms "good," "bad," "hot," and "cold" are independent of nature. These terms reveal the views and values of the person using them, not the physical process of creation or the inherent characteristics of people. Historically, many of the practitioners of eugenics viewed it as a science and not necessarily restricted to human populations; they embraced the views of Darwinism and social Darwinism.[327] Darwinists attempted to distance themselves from Hitler's practice of using their model of creation where the "fittest" survive. However, the model of "survival of the fittest" was used correctly. No one has shown where the theory was used incorrectly and differently from how it applied to animals. It is the model and its terms that allow almost limitless application to any living creatures, including humans. The fittest were determined by the reigning government in Germany, just as they were in the United States Supreme Court: the theory was the same and the words were the same.

Hitler: The Struggle for Survival—Sexual Selection

Erasmus Darwin used the concepts of struggle for survival and sexual selection; afterward, his grandson Charles used them. In the quote that follows, Hitler endorses them:

> The *struggle* for the daily livelihood leaves behind in the ruck everything that is *weak* or *diseased* or wavering; while the fight of the male to possess the female gives to the *strongest* the right, or at least, the possibility to propagate its kind. And this *struggle* is a means of furthering the health and powers of resistance in the species. Thus it is one of the causes underlying the process of development toward a higher quality of being.[328]

Hitler's endorsement uses the terms "weak," "diseased," and "wavering," which are part of Malthusian competition—the same competition embraced by Darwin. The sexual selection embraced by Hitler was also embraced by Charles Darwin's grandfather and later by Charles Darwin in *Origin of Species*.

Hitler: The Superior (Fitter) Race

If the terms used by Darwin to support his concept of natural selection are scientific, then those terms would also support Hitler's views as scientific (terms such as "inferior," "superior," "stronger," "weaker," "better," "degeneration," and "laws of race," among others). The same terms are used by Darwinists to this day. In *Mein Kampf*, Hitler writes about inferior and superior races, with the inferior always outnumbering the superior:

> The inferior would always increase more rapidly if they possessed the same capacities for survival and for the procreation of their kind; thus the best in quality would be forced to recede into the background. Therefore, a corrective measure in favour of the better quality must intervene.[329]

If nature does not wish for weaker individuals to mate with the stronger, she wishes even less that a superior race should intermingle with an inferior one, because in such a case all her efforts to establish an evolutionary higher stage of being throughout hundreds of thousands of years may thus be rendered futile. In short, miscegenation [mixing races] always results in the following:

(a) The level of the superior race becomes lowered.
(b) Physical and mental degeneration sets in, thus leading slowly but steadily toward a progressive drying up of the vital sap.

Hitler then writes much like Darwin:

He who would live must fight. He who does not wish to fight in this world, where permanent struggle is the law of life, has not the right to exist.[330]

This is right in line with Darwin's use of the same phrase. Darwin knew full well the deadly moral meaning of survival of the fittest and natural selection, as did Huxley: each person decided their own morality and there was no absolute arbiter, no God to determine otherwise. Darwin tried to soften what his readers thought. He attributed cruelty to the first part of natural selection, "chance," neglecting the second part of selection and direction that is supposed to nullify chance creation and provide a direction. He glossed over the fact that he opened the floodgates of moral arbitrariness. To soften the utterly stark coldness and the unimaginable atrocities that are accommodated by his theory, Darwin created a "bright face" for selection, in essence a mask. In *Origin of Species*, he writes:

When we reflect on this struggle, we may console ourselves with the full belief, that the war of nature is not incessant, that no fear is felt, that death is generally prompt, and that the vigorous, the healthy, and the happy survive and multiply.[331]

Hitler felt the same about the people in his gas chambers: "death is generally prompt, and that the vigorous, the healthy, and the happy survive and multiply."[332] Stalin felt the same about the millions of Ukrainians he starved to death. Chairman Mao's communists showed no concern for the countless numbers they murdered. Natural selection in the wild was cited as evil by Darwin with Ichneumonidae and pit bulls. But natural selection used by man was the worst, and it was the same model. It was magnified to catastrophic evil, but it operated under Darwin's "bright face" that "death is generally prompt, and that the vigorous, the healthy, and the happy survive and multiply." Darwin was educated and understood what he was doing, as do Darwinists today. Perhaps that is why Darwin's cousin, Galton, objected to "democracy," showing it is better to be in charge when such decisions are being made so as not to become a victim.

Interestingly, supporters of natural selection (i.e., survival of the fittest) rejected Hitler although Hitler was merely applying the very model they supported. There are no means of theoretically showing that any differences between Darwin's theory and Hitler's practice. For Darwin, man and animals were all one and the same. In Darwin's 1871 *Descent of Man*, he writes:

> [T]here is no fundamental difference between man and the higher mammals in their mental faculties.[333]

> It is notorious that man is constructed on the same general type or model with other mammals.[334]

If the models of natural selection and fitness are suitable for Darwinists, then the same words in the model would appear to be no less suitable for Hitler. The theory is the same. If natural selection were science, then the Nazi state would be a scientific state. Does one argue with science? In contrast, nothing was argued by Darwinists in the United States, where forced sterilization was first practiced well ahead of Hitler. It was the US Supreme Court that led the way for forced sterilization. If Hitler is a tyrant, would that also be true for the justices on the US Supreme Court and the Darwinists who fought for and supported that decision?

Racism was another outcome of natural selection. Charles Darwin's racism shows in his work *The Descent of Man and Selection in Relation to Sex*. Darwin reveals his indifference and that of his followers when he states that the superior races would destroy inferior races: one need not look further than the full title of his 1859 first edition of *Origin of Species*, which reads: *On the Origin of Species by Means of Natural Selection, or the Preservation of Favoured Races in the Struggle for Life*. Racism runs rampant in terms in which races are favored or unfavored. Wallace did not hold racist views, but Darwin was adamant about such views: In the *Descent of Man*, Darwin writes:

> At some future period, not very distant as measured by centuries, the civilized races of man will almost certainly exterminate and replace throughout the world the savage races.[335]

Who are or are not the savages depends on a point of view. Different governments' use of natural selection and survival of the fittest bears responsibility for the literal murder of many millions based on that view. Darwin showed a face of natural selection that he wished others to see, and he called it "the beautiful and harmonious diversity of nature."[336] These are supposed to be comforting words, but Darwin's fellow naturalists knew better: it was a diversity they wanted to "improve" by forced sterilization. Others used murder to "improve" society—with the model not indicating one way or the other. Even many amateur naturalists would have known better. Darwin himself, of course, knew better. In a letter to Hooker (July 13, 1856), for example, he writes of his views about nature and natural selection:

> What a book a Devil's chaplain might write on the clumsy, wasteful, blundering low & horridly cruel works of nature![337]

The cruel works of nature are the works of natural selection, according to Darwin. Darwin's softening of natural selection allowed his book to reach a wider audience, masking the harsh reality that would exist under natural selection—for people. Francis Galton, Darwin's cousin, wanted to "improve" society and that eugenics could be implemented with the "right" people at the helm of government. Galton writes about the type of government needed to improve society and it was not a republic such as that of the United States. In his view, a clear course of action was needed:

> It is the obvious course of intelligent men—and I venture to say it should be their religious duty—to advance in the direction whither Nature [not God] is determined they shall go, that is toward the improvement of the race...But it [democracy] goes farther than this, for it asserts than men are of equal value as social units, equally capable of voting, and the rest. This feeling is undeniably wrong and cannot last."[338]

Galton did not care for democracy. The United States, however, is not a democracy, but a constitutional republic. Yet, eugenics was first practiced in the United States with the use of forced sterilization. For approving forced sterilization, the name of Oliver Wendell Holmes, along with the other Supreme Court justices with whom he served, stands alongside Hitler's name in history. If Hitler was creating the Aryan Master Race with eugenics, what were the Darwinists and the US Supreme Court creating?

Summary

Survival of the fittest offers no means of determining right or wrong, as is also the case with natural selection. It only provides a license, an excuse, to justify whatever actions are taken or beliefs are held. There is only one inherent morality involved with survival of the fittest or with natural selection: the morality, such as it is, that comes from following one's feelings. Darwin and Hitler would agree on the words in the model, but perhaps not on the actions taken, *if* they had different feelings about those actions. Natural selection (survival of the fittest) created people and morality. The model's words do not change with the user, but the outcome of testimonial models, such as natural selection and survival of the fittest, do have different outcomes with different people using the same words. That is how testimonial models operate. For example, what is heavy for one person is light for another; what is cold for one is comfortable for another; what is evil for one person is good by another. Darwin could only offer his opinion on what he thought and felt was right or wrong, but he could not run away from the words he used and how they could be used by others. Based upon quotes from Darwin at the outset of this chapter, "multiply, vary, let the strongest live and the weakest die," he apparently embraces those words. In his 1871 book, *The Descent of Man*, Darwin writes:

> With savages, the weak in body or mind are soon eliminated; and those that survive commonly exhibit a vigorous state of health. We civilized men, on the other hand, do our utmost to check the process of elimination; we build asylums for the imbecile, the maimed, and the sick; we institute poor-laws; and our medical men exert their utmost skill to save the life of every one to the last moment.

> There is reason to believe that vaccination has preserved thousands, who from a weak constitution would formerly have succumbed to smallpox [discovered by Jenner before Darwin]. Thus the weak members of civilized societies propagate their kind. No one who has attended to the breeding of domestic animals will doubt that this must be highly injurious to the race of man.

> It is surprising how soon a want of care, or care wrongly directed, leads to the degeneration of a domestic race; but excepting in the case of man himself, hardly any one is so ignorant as to allow his worst animals to breed.[339]

139

As was correctly stated in the debate between the atheist Bertrand Russell and Fr. Copleston in the famous 1948 BBC radio debate (shown in "Chapter 4: Theistic Evolution") on the existence of God, during which the "Moral Argument" took place: without the absolute rules of the biblical God, moral vagary is determined by feelings. With natural selection or survival of the fittest as the creation model, feelings are the only way to determine morality and guide actions. This view essentially states there is no fixed morality; it is whatever a person wants it to be. Even for the same person, feelings change over time, thus changing their morality when they use natural selection or survival of the fittest.

In his 1888 book, *The Antichrist*, Friedrich Nietzsche (1844–1900), defines "good" and "evil" once it is broken away from the rules of the absolute morality. He writes:

> What is good?—All that heightens the feeling of power, the will to power, power itself in man. What is bad?—All that proceeds from weakness. What is happiness?—The feeling that power increases—that a resistance is overcome. Not contentment, but more power, not peace at all, but war; not virtue, but proficiency (virtue in the Renaissance style, virth, virtue free of moralic acid). The weak and ill-constituted shall perish: first principle of our philanthropy. And one shall help them to do so. What is more harmful than any vice?—Active sympathy for the ill-constituted and weak—Christianity.[340]

Feelings formed the very foundation of Nietzsche's morality. Nothing in Nietzsche's quoted view is out of context with natural selection or survival of the fittest. If power were increased, then the actions that led to that increase were "good," including genocide.

The role of natural selection and survival of the fittest was softened by Darwin so that it would be accepted, as it is softened by supporters of natural selection and survival of the fittest to this day. In the end, if power corrupts, *then natural selection ordains that corruption*. As long as natural selection or survival of the fittest use terms that are independent of nature, neither will ever constitute science. The same is true for evolutionary biology as well. As the terms of evolutionary biology can never be made dependent on nature's physical world, the fact that all models within it are composed entirely of terms that are independent of nature will render it a testimonial system—a religion, complete with its own morality—on an indefinite basis. There is no science in evolutionary biology.

CHAPTER 7

NEITHER NATURAL SELECTION NOR DARWIN IS NEEDED

Introduction: Biology Models That Successfully Portray the Physical World of Nature

In no case were Darwin's ideas, work, or model of natural selection needed by any man who worked in nature, including biology or science, either then or to this day. Giants in modern biology developed models that successfully operated in nature and as science, as shown by the diseases they cured and the successful discoveries they made. The sophistication of their research reached beyond Darwin's humble understanding and capabilities. They accomplished their achievements without the loose, independent language of Darwin's model of natural selection, which was content free of the physical world, its processes, rules, repetitiveness, and science. Louis Pasteur was Darwin's contemporary. Pasteur was doing his groundbreaking work in 1859, at the same time Darwin was writing his *Origin of Species*. Gregor Mendel also did his research into genetics and lived during Darwin's time. Mendel's work later necessitated the formation of the second "Darwinian synthesis" of the 1930s and 1940s. Neither Darwin nor natural selection helped develop these men's highly successful models of nature that saved lives using germ theory (still new at that time), and advancing biology's models that existed in nature. The men who advanced biological science appear in the following list, including those who lived as Darwin's contemporaries.

143

1. Edward Jenner (1749–1823): English physician
2. Robert Koch (1843–1910): bacteriologist
3. Louis Pasteur (1822–1895): chemist and biologist
4. Gregor Mendel (1822–1884): geneticist
5. Martinus Beijerinck (1851–1931): microbiologist

A short description of each of these men's major accomplishments follows, along with a description of how their work was independent from Darwin: neither they nor anyone else had any need of Darwin or natural selection.

1. Edward Jenner (1749–1823): English physician

The English physician Edward Jenner performed experiments, beginning in 1796 with the vaccination of eight-year-old James Phipps, which proved that cowpox provided immunity against smallpox. He did this sixty-three years before Darwin published *Origin of Species*. Jenner's discovery was instrumental in ridding the world of smallpox and laid the foundation for modern immunology.[341] He did not need or use natural selection or evolutionary biology in any of his work.

2. Robert Koch (1843–1910): bacteriologist

Robert Koch won the Nobel Prize for physiology or medicine in his investigations and discoveries in relation to tuberculosis in 1905. He received numerous other medals and honors during his lifetime and after his death. He is credited with developing many innovative laboratory techniques and proving that microorganisms (germs) cause anthrax, cholera, and tuberculosis. His work was essential in proving the germ theory of disease and the contagion of such diseases. Koch also proved instrumental in applying the germ theory to public health and hygiene practices in order to prevent disease in his native Germany and elsewhere.[342] He did not need any evolutionary biology models.

3. Louis Pasteur (1822–1895): chemist and biologist

By 1857, two years before Darwin published *Origin of Species*, Pasteur had become world famous, and he took up an appointment as director of scientific studies at the École Normale in Paris. Pasteur was asked to help to investigate a serious disease that was ruining the silk industry in southern France. The disease, known as pébrine, attacked the silkworms.[343] The signs of the disease were either that the eggs did not hatch or that the worms would die before making their silk cocoons. The disease had reached epidemic proportions, and even disease-free worms brought in from Spain and Italy had become contaminated. By 1864, no uncontaminated eggs remained except for those brought in from Japan.[344] Pasteur then worked with the silk industry to devise a simple way to keep the worms disease-free.[345]

Not only did Pasteur rescue the French silk industry, he also established the causal link (the connection) between bacteria and disease. That link had not been fully understood previously, which means that this was a major discovery.[346] The connection between the bacteria (the cause) and disease (the effect) formed the "germ theory" model used by Pasteur. The ancient historical view of disease held that disease occurred through spontaneous generation instead of developing from microorganisms that grow by reproduction.[347] The germ theory of disease is also called the "pathogenic theory of medicine." This model of nature linked microorganisms to the cause of disease, with specific microorganisms causing specific diseases. Although highly controversial when first proposed, the germ theory now forms a scientific cornerstone of modern medicine and clinical microbiology.[348] Germ theory and development of vaccinations did not require any use evolutionary biology models of creation.

3a. Pasteur's Vaccination Model Cures Anthrax

In Pasteur's France, from 1822 to 1895 (Darwin died in 1882), many cattle suffered from anthrax, a serious disease that killed many. Pasteur made a careful study of anthrax and its physical effects and noticed that some cows developed the disease more severely than others. So he decided to inject two cows with a strong dose of the anthrax bacteria, fully expecting them to die. To Pasteur's amazement, neither of them developed the disease. Later, he found that both animals had already suffered from anthrax. Could they be immune to it? Could they be protected in some other way? Pasteur believed that if he could give an animal a mild attack, this might be sufficient to prevent it from getting the disease later.[349]

Eventually, after many physical experiments using the physical parts of nature (anthrax germs and cows), Pasteur succeeded in producing a weakened, harmless culture of anthrax bacteria. He used this to inoculate cattle and sheep, giving them a mild form, which they recovered from. When these animals mingled with others that had a severe form, they remained unaffected. They were immune, illustrating the operations of Pasteur's vaccination model.[350] The application of Pasteur's vaccination model had worked successfully. Pasteur worked throughout the rest of his life on the various causes of disease and how these could be prevented by vaccination.[351] The research performed by Darwin paled in comparison to that performed by Pasteur for anthrax. When reading about Pasteur's exploratory work, his identification of diseases, and his tests for the effectiveness of his vaccines, it is plain that his work never involved any claims of evolution or creation. None of the terms or models used in evolution—"evolutionary biology," "creation," "fitness," or "adaptations"—was ever used.

3b. Pasteur's Vaccination Model Cures Rabies

Pasteur is particularly renowned for his work on the rabies vaccine. Rabies is a highly contagious infection that attacks the central nervous system. It enters the body through the bite of an infected animal or through infected saliva entering an existing wound. After experimenting with the saliva of animals suffering from the disease, Pasteur concluded that the disease rests in the central nervous system of the body. When an extract from the spinal column of a rabid dog was injected into healthy animals, it produced symptoms of rabies. By studying the tissues of infected animals, Pasteur was able to produce a weakened form of the virus that he could use for vaccination.[352]

On July 6, 1885, three years after Darwin's death, Pasteur tested his pioneering rabies vaccine for the first time on Joseph Meister at the urging of others. He saved the young man, who had been bitten by a rabid dog. The treatment lasted ten days, and at the end the man recovered and remained healthy. Since then, countless people have been treated and saved by Pasteur's vaccination model for rabies.[353]

In March 1886, Pasteur was invited to present his results to the Academy of Sciences. In 1888, he went on to found the Pasteur Institute in Paris, which was a pioneering clinic for the study of infectious diseases and the treatment of rabies and a center for teaching. Pasteur directed the institute personally until he died. The Pasteur Institute is still one of the most important centers in the world[354] for the study of biology, microorganisms, vaccines, and infectious diseases. It was the first to isolate HIV, the virus that causes AIDS, in 1983. It has been responsible for breakthrough discoveries that have enabled control of such virulent diseases as diphtheria, tetanus, tuberculosis, poliomyelitis, influenza, yellow fever, and plague. Since 1908, eight Pasteur Institute scientists have been awarded the Nobel Prize for medicine and physiology, and the 2008 Nobel Prize in physiology or medicine was shared with two Pasteur scientists.[355]

Pasteur became a national hero and was honored in many ways. He died at Saint Cloud on September 28, 1895, and was given a state funeral at the Cathedral of Notre Dame; his body rests in a permanent crypt at the Pasteur Institute.[356]

Darwin's model of natural selection does not apply to any work with germs or vaccination; it is not physically used in curing any diseases, for it cannot be used to physically relate "cause to effect"; it has no basis in nature for relating to a single fossil, which it supposedly created.

3c. Pasteur Shows the Spontaneous Generation Model to Be False

In Pasteur's time, people commonly believed that "life arose from nonlife." Even disease was thought to be caused by spontaneous generation.[357] In one view, mice were created by grains of wheat and a dirty shirt because when the two were placed together, mice were found. In another version, mice were also believed to come from dirty underwear. Scientists once thought that maggots came from rotting meat. In that time, there even existed a spontaneous generation causal model for creating mice from nonlife, which was to take sweaty rags, wrap them around wheat, and set them in an open jar. In twenty-one

147

days, according to the model, you would "create" mice.[358]. All of these ideas began to fade in 1859 with Louis Pasteur's experiments dealing with spontaneous generation. With his experimental accomplishments, Pasteur concluded in 1864,[359] "There is no known circumstance in which it can be confirmed that microscopic beings came into the world without germs, without parents similar to themselves." Louis Pasteur demonstrated that "life comes only from life," a view still accurate today. Pasteur had set the stage for modern biology and biochemistry.[360] In sharp contrast to Pasteur, Darwin's views on spontaneous generation proved vague and uninformed through experimental evidence.[361] Spontaneous generation has not been shown to have a credible causal model basis in nature.

4. Gregor Mendel (1822–1884): geneticist

In 1866, seven years after Darwin published the 1859 *Origin of Species,* Gregor Mendel published his paper, "Experiments on Plant Hybrids," in the *Proceedings of the Natural History Society of Brünn.*[362] It was Gregor Mendel, an Austrian Augustinian priest,[363] who gave us the model of heredity held to this day, for which he is known as the father of modern genetics. Between 1856 and 1863, Mendel cultivated and tested approximately 29,000 pea plants (i.e., *Pisum sativum*). His experiments brought forth two generalizations that later became known as Mendel's laws of inheritance. Mendel did not need or use natural selection. He neither needed nor used any of Darwin's work. Gregor Mendel believed that his discoveries refuted Darwin's premises about the heritability of traits, which he called pangenesis.[364] Pangenesis holds that body cells shed gemmules, which collect in the reproductive organs prior to fertilization. Thus, every cell in the body has a "vote" in the constitution of the offspring. The skipping of generations with the recurrence of ancestral features, called atavisms, is said to arise due to the awaking of long dormant gemmules, while limbs regenerate due to the activation of gemmules from the missing limb.

Modern genetics began with the work of Mendel, whose breeding experiments with garden peas led him to formulate the basic laws of heredity.[365] Mendel's work formed the seminal foundation of genetics and heredity that ultimately led to the knowledge of DNA and the human genome. He published his experimental findings in 1866, but his discoveries lay unknown for thirty-four years until 1900, when a number of researchers independently rediscovered Mendel's work and grasped its significance.

Mendel came to three important reproducible conclusions from these experimental results:[366] that the inheritance of each trait is determined by "units" or "factors" (now called genes) that pass on to descendants unchanged;[367] that an individual inherits one such unit from each parent for each trait;[368] and that a trait may not show up in an individual but may still pass on to the next generation (skipping a generation, called "atavism").[369] In contrast, Darwin's theory of inheritance, called "pangenesis,"[370] never succeeded. Mendel's research spanned decades of observable progress. As such, that research formed the basis of modern genetics, an achievement accomplished without any evolutionary models.

5. Martinus Beijerinck (1851–1931): microbiologist

Martinus Beijerinck, a Dutch microbiologist, was the first person to use the term "virus" to label the invisible, disease-causing organisms he showed to be self-replicating. He originated selective culture techniques, also known as enrichment culturing, and was the first to isolate a wide range of microorganisms.[371] Beijerinck made major contributions to microbiology by developing the enrichment culture technique simultaneously with Sergey Winogradsky, which permits the isolation of highly specialized microorganisms. In studying tobacco mosaic disease, Beijerinck identified the filterable pathogen as a "contagium vivum fluidum," a term he coined to convey his concept of a living, infectious agent in a fluid (noncellular) form—a revolutionary idea at a time when life and cellularity were thought to be inextricably connected.[372] His work was rooted in research dramatically more sophisticated than Darwin's, whose evolutionary theories and claims went unused by Beijerinck.

Summary of Natural Selection and Darwin Not Being Needed

Researchers developed and used biological models of nature to cure smallpox, rabies, polio, and other crippling and deadly diseases. They did not need

Darwin or natural selection. These researchers' foretelling science models ushered in and advanced modern biology without the use of natural selection or evolutionary biology. Evolution by natural selection, or just evolution alone with no associated cause, has never shown accomplishments that merited thinking of it as being "at the core of genetics, biochemistry, neurobiology, physiology, ecology, and other biological disciplines, as claimed by the NAS."[373] It is no wonder that notable biologists did not use natural selection. They could not, even if they wanted to do so. As Australian Hiram Caton writes:

> Pasteur didn't include evolutionary factors in his research because he didn't deem them to be relevant. He wrote nothing about evolution and did not include an evolutionary component in the Pasteur Institute. A similar story may be told of other eminent experimentalists of that time. Germany's leading cell pathologist and doyen of medical science, Rudolph Virchow, opposed Ernst Haeckel's attempt to Darwinize biological science. The same is true of embryologist Wilhelm His, who strongly criticized Haeckel's recapitulation theory. In France, neurologist Paul Broca and anatomist Claude Bernard fit a similar pattern.[374]

The German Virchow denied that natural selection had any legitimacy in nature;[375] he was far from alone. Natural selection could no more be used in any operations of nature than could miracles or special creations as they each operate through testimony linking cause to effect. Were it not for those men that gave us the foretelling models of biology that are independently repeatable, the world as we know would have health difficulties that burdened people with despair and hardship. At the core of biological cures and discoveries are these men and their foretelling biological models, without which disease would begin to sweep into human and animal populations again with diphtheria, polio, tetanus, smallpox, rabies, cholera, anthrax, measles, and other deadly and disabling diseases. It is modern, not evolutionary, models of biology that enable cures and preventions of diseases; modern biology is without the taint of evolution by natural selection. It is the foretelling models, not evolutionary models, that relate cause to effect: vaccinations the cause and cure of a specific disease is the effect. Evolutionary models are testimonial and show no cause and effect linked by nature's physical world.

CHAPTER 8

THE CHARACTERISTICS OF THE MODELS CALLED "SCIENCE"

Introduction

Science models are like recipes; they contain everything needed to have "cause and effect" take place: each model names all the parts of nature that are involved, it describes the relationships of each part to the others, it identifies quantities of each part in the model, it shows proportions, it provides the rules that govern the operations, and it shows how cause produces effect. Finally, the model foretells the effect to be observed, which is then confirmed by repetitive independent observations. The science model's identifying characteristics do not depend on the model's method of conception, its means of discovery, or its subject matter (physics, biology, chemistry, magnetism, and genetics), or its endorsements or popularity. An example of the unimportance of any method used in developing a science model is Newton's model of gravity, which is used today with no need to know the method he used to develop the models, the data he gathered, the age of the earth, or the Christian beliefs he held about creation. Newton's models operate successfully without knowing their history or the manner by which it was conceived. Models that are science characteristically stand independently of any method, such as the scientific method. It is the successful foretelling of the model that determines its being science.

The reason science models are prized above other types is because they can be used to create things that are desirable. Technology is created in this way. It is the foretelling models that have gained public recognition and desirability and are called "science." Vaccinations, radios, televisions, and telephones are examples of the benefits of using science models for the creation of technology. The combination of each model's foretelling, followed by confirming observations, shows the model to represent nature and have the characteristics that show them to be science. If models' of nature and science could only be spoken, like holy pronouncements, they would be just like evolutionary biology models, which are incapable of being used for anything other than belief and faith: they would have no role in nature's physical world.

Science's Characteristics

A brief description of the characteristics of science follows.

1. Science's Characteristic: Foretelling.

Over 2,300 years ago, the Greek mathematician Archimedes (c. 287 BC–c. 212 BC) described how nature's processes operated by developing models that are used to this day to accurately portray processes and foretell (predict) outcomes. Archimedes is often considered to be the greatest mathematician of antiquity—and one of the greatest of all time. For instance, he calculated the area under the arc of a parabola and accurately computed the value of a circle's circumference divided by its diameter, which is used so often that it was given a name: the Greek letter *pi*, whose accurate approximation is 3.14159. He also defined the spiral bearing his name and was a physicist and engineer. He discovered the model that shows how to balance the lever— essentially a seesaw. He created siege engines and developed a screw pump.[376] One of Archimedes' models could foretell the upward buoyancy force exerted on an object when placed in water or a balloon in the atmosphere. That model, called the "buoyancy model," still operates successfully over 2,300 years later. As long as nature operates in the same manner, the model will always operate accurately.

With a successful foretelling model of nature, one can apply the models to harness nature. With this type of model, technology can be created, such as: vaccinations that prevent or cure illnesses; identification and manipulation of genes; aircraft that enables man to fly; televisions that can receive signals and show images on a screen; radios that send speech through space; machines that can generate and use electricity; engines that can move vehicles, pump fluids, or sail ships; and electronic circuits that change the way people live. The ability of a model to tell us what is going to take place is critical to designing technology that works in nature. Without a model that successfully foretells the outcome of nature's processes, there could be no engineering, modern medicine, modern biology, or modern science.

2. Science's Characteristic: Confirming Observations

A cause and effect model that renders a foretelling (makes a prediction), which is followed by observations that match the foretelling, and is independently repeatable, is a model that is called science. As foretelling models do not provide for testimony, they never conflict with a person's worldview or religion. There can be no conflict between a science model and religion. Religion is testimonial. Different people may use a foretelling model at different times and still find that the model's foretellings match confirming observations. The more times that a model foretells what is to be observed and is followed by confirming observations, the more it is shown to operate successfully in nature, allowing it to be called "science." Newton's model of gravity foretells the amount of pulling force exerted between two bodies, such as a satellite and a planet or a projectile and its path. The model of gravity was used by Newton's friend Edmund Halley when he observed a comet, used the model, which successfully foretold when it would be observed again. That comet's reappearance confirmed the foretelling of the model of gravity and was named "Halley's Comet." Each successful, subsequent observation confirms that the model represents the operations of nature. Even the discovery of new planets (Uranus, Pluto), first thought to discredit the model of gravity, ultimately confirmed it.

Pasteur's vaccination model foretells a cure or prevention of a specific disease. The successful cures and preventions provide the confirming observations for Pasteur's vaccination models (rabies, anthrax, cholera). In another example, Mendel's genetic model foretells that a pattern of physical characteristics will be revealed, much like a pattern of outcomes in any one game of chance (with a known population). Faraday's electromagnetic induction model foretells that "whenever a magnetic force increases or decreases, it produces electricity, the faster it increases or decreases, the more electricity it produces."[377] Independent observations confirm this model as well as each of its physical parts. Successful models of nature foretell what is to take place in nature and show each part of nature that is involved in the model.

In Einstein's general relativity theory model, mass "warps" space and time to create gravitational fields and bend light. During a solar eclipse in 1919, Einstein's model was confirmed when Arthur Eddington observed that light from stars passing close to the eclipsed sun was bent, so that the stars appeared slightly out of position, thus confirming the model's foretelling. Models that do not or cannot foretell what will be observed do not portray nature accurately, if at all.

3. Science's Characteristic: Independent Repeatability

Repeatability means that a model of nature may be used by different people and still obtain a model's foretelling that is matched by confirming observations. A model that is not independently repeatable produces different results for different people and is not useful in revealing nature's operations. Engineering may only be performed using models that possess the capability of repeatedly rendering accurate foretellings that are then followed by confirming observations. Each use of a science model confirms its legitimacy of accurately representing nature's processes. This was true for the models of Archimedes, Copernicus, Newton, Pasteur, Mendel, and every model that is science, but *none* in evolutionary biology. With repeatability, Newton's model of gravity can be used to compute the amount of force foretold by the model, to be followed by confirming observations. Any person can compute the force of the earth acting on a satellite, followed by the model showing how rapidly the satellite must orbit the earth to avoid going off into space or crashing back down. Different people will arrive at the same answer using the model of gravity and the models of motion. The launching of satellites into orbit shows the foretelling and confirming observations.

Biological repeatability must exist for a model to possess the characteristics of science. Pasteur's vaccination model for anthrax, rabies, and cholera proved repeatable for anyone using them. Anyone may achieve the same results using models that are science. Repeatability does not depend on a person's acceptance of the model or the user's religion. One may even reject a successful repeatable model as being false, use it, and still find accurate foretelling to match confirming observations. The model operates in the same way for every person, regardless of his or her beliefs, because science models are independently repeatable—unlike those in evolutionary biology, such as natural selection, selection pressure, arms races, and adaptive landscapes.

Examples of Models That Possess the Characteristics of Science

Here follows a brief description of several models that possess the characteristics of science. These models are Archimedes' lever model, Newton's model of gravity, Mendel's genetic model, and Pasteur's vaccination model.

1. Archimedes' Lever Model

Over 2,200 years ago, hundreds of years before the birth of Jesus Archimedes created a model of the bar and fulcrum. This is known as the lever model and is illustrated by the simple seesaw that children play on. All of the lever model's components exist and operate in nature's physical world, unlike the models of evolutionary biology. Archimedes' lever model states:

> Two forces acting in opposite directions will balance when
> the product of one force and its distance from the pivot point
> is equal to the product of the other force and its distance from
> the pivot point.

In other words, the forces of a seesaw balance when the force times the distance on one side of the pivot point equals the force times the distance on the other side. This model shows how to use lever action to move a heavy object with very little force. Many students study this simple model in high school, and freshman engineering students also review it. Using this model, a person could foretell the force needed to "move the world,"[378] as Archimedes put it. This model shows all the operative parts, their relationships, and the rules that govern them in the "cause and effect" model. The model may be written using words or mathematics. All the parts of the model exist in nature and remain the same for every person using it. Knowing three of the four parts of the model, the model can foretell the fourth. Notice that there is no place for testimony, for the model is self-contained. All the terms in the model are dependent on nature. That is, they are observable. These terms are "forces," "distances," and "direction" (up, down, right, and left).

Archimedes' lever model is useful because it foretells the amount of force that must be applied to one end of the lever to lift an object at the other end, like a seesaw[379] (or teeter-totter) or a lever, thereby accurately portraying how nature operates. We can test this simple model by taking its foretold force and applying it to a fulcrum or seesaw. If the seesaw comes into balance, then the foretelling of the model matches the observed balancing. The characteristics of science that are possessed by the model are foretelling, confirming observations, and independent repeatability of the model's results.

2. Newton's Model of Gravity

Newton's law of gravity states: "Every object attracts every other object, by virtue of their having mass. An object with twice the mass will attract other objects with twice the force. An object twice as far away will attract other objects with one-fourth the force." In other words:

> Every particle of matter attracts every other particle of matter with a force directly proportional to the product of the masses and inversely proportional to the square of the distance between them.[380]

This is Newton's model of gravity. It may be written in the form of an equation, but the underlying meaning of the words or the equation remains the same. When stated using mathematics, it can easily be used to easily solve problems, which is the advantage of mathematics. When Newton introduced the gravity model, no one knew how gravity operated. People did not "debate gravity,"[381] as has been claimed; they debated different "cause and effect" models of gravity, differing on which one actually portrayed gravity's operation in nature. Gould writes, "Einstein's theory of gravitation replaced Newton's," which is an inaccurate statement. Newton's model of gravity operates with extremely high accuracy at speeds well under that of light. Approaching the speed of light, Newton's model of gravity does not give accurate foretelling when compared with observations. When Newton introduced his gravity model, the absence of a mechanical push or pull was seen as mystical at the time as the prevailing worldview was mechanical. But that did not change the fact that the model operated successfully by foretelling observations. The model overcame all objections by its consistent ability to foretell what would take place and then have the foretelling match observations, which confirmed it.

The English astronomer Edmund Halley was a good friend of Isaac Newton. In 1705, he used Newton's new theory of gravitation to determine the orbits of comets from their recorded positions in the sky as a function of time. He found that the bright comets of 1531, 1607, and 1682 had almost the same orbits; and when he accounted for the gravitational perturbation on the comet's orbits from Jupiter and Saturn, he concluded that these represented different appearances of the same comet.[382] He then used Newton's model to foretell the return of this comet in 1758. Seventy-six years after Halley used the gravity model's foretelling, the comet appeared again, and the model was thus validated by the first confirming observation.[383] Because of the model's success, the comet was first named in Halley's honor by French astronomer Nicolas Louis de Lacaille in 1759.[384] Thus, Newton's model of gravity became science.

The confirmation of the comet's return represented the first time anything other than planets had been shown to orbit the Sun. It also served as one of the earliest successful tests of Newtonian physics and a clear demonstration of its foretelling explanatory power in nature, using the parts of nature, their physical relationships, and the model's foretelling of cause and effect. The physical parts of Newton's model of gravity include force, mass, distance, and a constant. Some of Newton's friends offer an excellent illustration of how a model with the characteristics of science operates. Nicolas Fatio de Duillier and David Gregory write, "Newton thought that gravitation is based directly on the will of God."[385] The fact is, however, that the model of gravity operated in the same manner, with the same results, without reference to presence or absence of God in the model and without any change of beliefs in God on the part of the user. An atheist or a Young Earth creationist could use the model and arrive at the same results. That is a fact because the models of religion and science do not conflict. That cannot be said for natural selection, which necessitates that the user, if a believer in Genesis, change "how to think" about the world and adopt a new belief system.

3. Gregor Mendel's Genetic Model

Gregor Mendel was a contemporary of Darwin, as was Pasteur. Mendel developed a model that successfully foretells the characteristics of parents' offspring over generations. Mendel opened the door to finding the link between genetics and the body, showing that no new changes take place. Mendel was inspired by both his professors at Olmatz University and his colleagues at the monastery to study variation in plants, and he conducted his study in the monastery's garden. Between 1856 and 1863, Mendel cultivated and tested some 29,000 pea plants (i.e., *Pisum sativum*). This study showed that one in four pea plants had purebred, recessive alleles, two out of four were hybrid, and one out of four was purebred dominant. His experiments brought forth two generalizations that later became known as Mendel's laws of inheritance. Mendel observed seven easily recognized traits in the pea plants that only occurred in one of two forms:[386] (1) flower color is purple or white; (2) flower position is axil or terminal; (3) stem length is long or short; (4) seed shape is round or wrinkled; (5) seed color is yellow or green; (6) pod shape is inflated or constricted; (7) pod color is yellow or green. Using these characteristics, Mendel showed that the exact outcome of any single trial using his genetic model was a matter of "chance," with all possible outcomes forming the total population from which chance could be computed. The pattern of outcomes was physically identifiable. The "chance" or probability of any one outcome was known beforehand, just as in all games of chance. Natural selection's chance operated with no existing population, showing it to be an emotive type of chance—one not existing in nature.

4. Pasteur's Vaccination Model

Anthrax is not only one of the oldest known diseases, but its causative agent was also the first in history to be definitively isolated, reproduced, and linked to the disease that affects both livestock and humans.[387] One researcher, Robert Koch (1843–1910), observed and even photographed the results of the anthrax disease over a long period. Pasteur (a contemporary of Darwin's) would draw upon Koch's artificially grown cultures to develop a vaccine for anthrax in 1881.

Using the vaccine model, Pasteur used the same methods to devise a rabies vaccine three years later.[388] Pasteur composed the vaccine of specifically identified Anthrax germs, showing that germs caused the disease—although at that time, some thought that germs did not cause diseases.[389] Pasteur proved them wrong with his tests using the vaccination model. In one test, Pasteur used weakened cholera germs as a vaccine for chickens.[390] He then gave the chickens full-strength germs and found the chickens were immune to the disease. A new worldview and vaccination model was born where germs caused disease and a specific germ caused a specific disease. That vaccination model included using weakened germs to provide immunity to that specific disease. Over time, Pasteur developed immunization models for cholera, anthrax, and rabies. The immunization models both prevented and cured diseases. Weakened germs[391] served as a key component of the vaccine model.

When reading about the many experiments conducted by Jenner, Koch, Pasteur, and others, it becomes readily apparent that evolution neither played a part in the research nor was related to that research, nor did the cause of evolution enter into the research or vaccination models developed to combat diseases.

Summary: The Meaning of Modern Science

Science and religion have characteristics that identify each of them. Those characteristics show one to be different from the other, regardless of the number of tests that are conducted. The tests show science to be different from religion. Science is a particular type of model that portrays nature's processes of cause and effect that are linked by nature, not testimony. Science models possess the characteristics of foretelling what is to be observed, followed by independent observations that confirm the model's foretelling. Independent repeatability is the third characteristic of science models. Without science models, there would be no lifesaving vaccinations, no radios, no televisions, no satellites, no electronics, no space travel, and no modern life.

Evolutionary biology models' characteristics are radically different from those of models that are science; their characteristics are easy to identify, such as not having physical characteristics. Evolutionary models possess the characteristics of testimonial faith-based models, such as those used in the major religions. One of the characteristics that differentiate testimonial models is that their components are not dependent on nature: they have no measurability to show an existence in the physical world. That is, they do not operate in nature, but by testimony, as is the case for populationless chance, good variations, bad variations, accumulations, selection, direction, fitness, and adaptation. Natural selection and all evolutionary biology models use testimonial models where nature's "cause and effect" is replaced by "testimony and effect." Examples of "testimony and effect" models are "inference and effect," "correlation and effect," and "extrapolation and effect." *Both types of models, science and testimonial, use testing.* The testing is called "foretelling testing" for science and "testimonial testing" for evolutionary biology and religion. Engineering and science models, where it is understood that only physical components of nature are used, only use the term "testing." The two types of tests are dramatically different. Science models operate in nature's physical world, which is their fourth and final characteristic. Science models show how nature can be manipulated for man's applications. Testimonial tests, which are the only ones that may be used in evolutionary biology, operate solely within a person's mind, operating to support "how to think about" and relate to the physical world. Testimonial tests are illustrated in chapter 11, "Testing Miracles and Natural Selection."

Of the two types of models—science and testimonial—it is the testimonial models that are most important. Testimonial models are the source of morality; they are the basis for a person's actions, making them more important. The *use* of science models is determined by the application of testimonial models. One example of a testimonial model is the morality given by the Ten Commandments. Another example of testimonial morality derived from feelings is the cliché, "If it feels good, do it." In general, the morals that come from testimonial models have two sources: an absolute source (from God) and from feelings (arbitrary and possibly different from each person). Morals come from testimonial models found in Genesis and the rest of the Old and New Testaments. Evolutionary biology morality comes from feelings, as discussed in *Chapter 4: Theistic Evolution*. It is only testimonial models that determine the moral actions of men and women. The two types of testimonial morality—absolute and evolutionary biology—are of vast importance because they result in morality that determines the behaviors of people and the nations they form. Examples of nations formed by the use of morality are Russia, Nazi Germany, China, and the United States. There is a reason that people have flocked to the United States and have fled from Russia, China, and Nazi Germany: few wanted to live under a morality determined by Darwin, natural selection, and survival of the fittest.

CHAPTER 9

THE MEANING OF VARIATIONS

I believe that one day the Darwinian myth will be ranked the greatest deceit in the history of science.
—Soren Lovtrup, *Darwinism: The Refutation of a Myth,* 1987

- Professor of Embryology, Dept. of Animal Physiology at the University of Umea Sweden.
- Chairman of the Organization of Swedish Developmental Biologists, SDB, from 1979 to 1987; served as the first chairman.

Variations: Natural Selection's "Raw Materials"

Those hoping to find the definitive meaning of the term "variation" in the model of natural selection will forever be disappointed: the definition will not be found there, in nature, or anywhere else. Although there are fixed meanings for terms used in science models (such as meters, light-years, coulombs, amperes, pounds, rabies germs), the meanings given to "variation" are not fixed or known ahead of time. Rather, they are given an endless parade of meanings that are attributed on a case-by-case basis, as if taken from some ethereal realm where definitions are arbitrary and unanchored to any physical reality. An example of a term that is not anchored to a physical reality is "heavy." It cannot be measured. There are no units by which it can be identified. It is not dependent on nature, and it varies with each use—and each

165

person using it: what is heavy for one person may not be heavy for another. In comparison, the term "thirty pounds" is dependent on nature and is the same for each person. What is thirty pounds for one person is thirty pounds for another person, but what is "heavy" for one person may not be "heavy" for another. Dependency on nature is a characteristic of the terms used in science, but not "evolutionary science" or "evolutionary biology," which thrive on terms like "heavy," "good," "bad," "rapid," and "gradual."

The process of defining "variations" changes with each different case that is encountered. Some of the meanings of variations signify changes in the length of an animal's nose or beak, the proportion of black and white colors of a moth population, the speed of a race horse, the weight of the animal (such as a pig), the amount of wool from a sheep, the volume of milk from a cow, or shapes of feathers on a pigeon. When variations are said to accumulate, the claim is likely to be incoherent. It could mean pounds being added to quarts, or muscle being added to blood, or chemicals accumulating with eye parts—such variations make no sense in a physical reality, but they are the basis of Darwin's work, natural selection, and survival of the fittest. The user of the terms must give the terms consistency and meaning that does not exist in the model or in nature. There are many types of variations that can be used in "evolution by natural selection," none of them dependent on nature. The variation's *meaning by attribution* necessitates a case-by-case redefinition of the term; otherwise, the incoherency becomes immediately apparent. Neither Darwin's model nor Darwin ever showed how variations and accumulations operated in nature. Due to the lack of a fixed definition of "variations" that was dependent on nature, one knows no more about what creation steps took place after using "variation" or "accumulation" than before using the term.

The lack of a fixed definition for "variation" should have bothered Darwin, as it should trouble users of the term today. However, that does not seem to be the case, despite the fact that the absence of the definition is fatal to a model that is said to operate in nature and science. For terms that are dependent on nature, the terms are fixed and mean the same thing for every person, such for pounds (weight), miles or kilometers (distance), quarts or liters (volume), amperes (electricity), or bacteria (specific germs causing illness), each of which is dependent on nature. Dependency on nature is critical to models that show how nature operates as well as showing the models are science. Without that dependency, a model of science cannot exist; yet none of the key terms in evolutionary biology is dependent on nature, removing those terms and evolutionary biology from being used in science for all time.

Importance of Variations

In his 1889 book, *Darwinism,* A. R. Wallace emphasizes the importance of variations when he writes:

> The foundation of the Darwinian Theory is the *variability* of species, and it is quite useless to attempt even to understand that theory, much less to appreciate the completeness of the proof of it, unless we first obtain a clear conception of the nature and extent of this variability.[392]

Variations took on an importance for Darwin and Wallace when each man read Reverend Thomas Robert Malthus's book, *An Essay on the Principle of Population* (1826, sixth edition). In his autobiography, Darwin writes about Malthus's *Population* and its effect on him, including his view of variations:

> In October 1838, that is, fifteen months after I had begun my systematic enquiry, I happened to read for amusement Malthus on *Population,* and being well prepared to appreciate the struggle for existence which everywhere goes on from long-continued observation of the habits of animals and plants, it at once struck me that under these circumstances favourable [good] variations would tend to be preserved, and unfavourable [bad] ones to be destroyed. The result of this would be the formation of new species. Here, then, I had at last got a theory by which to work;[393]

For Darwin, the theory "by which to work" would become:

> [A]ll organic beings have been developed through variation and natural selection.[394]

This is the gradual model of creation that forms evolution, at least in Darwin's mind. Darwin writes it this way:

> Then, from the many slight successive steps of variation... The fore-limbs, for instance, which served as legs in the parent-species, may become, by a long course of modification, adapted in one descendant to act as hands, in another as paddles, in another as wings...[395]

According to Darwin, variations are important because they gradually accumulate to change forelimbs into hands, paddles, and wings. Variations change single cells into elephants and bears into whales, if you believe Darwin. Knowing about these changes that are said to take place, it is easy to see that without variations, nothing can be created by natural selection. The major flaw in these claims is that no accumulations of variations were ever observed to take place.

Variations Physical Characteristics: The Immovable Locations

Natural selection does not reveal how the thousands of "good" eye variations are joined together "in a direction" to form the eye. Darwin's work never describes how selection identified good variations that are accumulated for each different organ in each different creature. Darwin merely writes that good variations accumulated:

> This preservation of favourable variations and the rejection of injurious variations, I call Natural Selection. Variations neither useful nor injurious would not be affected by natural selection...we may feel sure that any variation in the least degree injurious would be rigidly destroyed.[396]

One significant point that is not addressed by Darwin is that once a variation is created by chance (good or bad), it cannot be moved, discarded, or destroyed, despite Darwin's claim that "any variation in the least degree injurious [bad] would be rigidly destroyed."[397] He never describes how that destruction would be performed by natural selection. Neither Darwin nor anyone else addresses how "bad" variations could be removed, possibly because there are no means of removing harmful or injurious variations. Once created, each variation is physically part of an accumulation, also called a "modification," which is part of a body that is anchored in some chance determined location for all time. The variations are the "raw material" from which new creatures are created; they are generated randomly in the first of natural selection's two parts. The location of each variation is in a fixed random location and cannot be moved afterward. Nor can the composition of the variation be changed, however unsuitable it is (note that it is always unsuitable). Created randomly, for example, a heart variation need not be placed in the heart but rather could be placed anywhere in the body, begging the question: how does the completed heart get formed when its variations are scattered throughout the body? A heart variation may be incomplete, made of the inappropriate material, or not match the rest of the heart. Consider the case where a heart variation may be created once and never created again. Natural selection does not show how any creation could ever take place, such as the creation of the heart where specific variations are needed; if the necessary variations are created at all, they may be scattered in many bodily locations, preventing them from being accumulated to form the heart. Specific materials, shapes, dimensions, locations, and so on, are needed for completion and without these very specific variations, the heart will never be successfully created. What is true for the heart is true for every part of the body of every creature.

Variations Physical Characteristics: Operations

No organ is capable of operating all by itself. When a single organ is created, it needs others to operate. Each organ is part of a system, and each system requires other systems to operate in the body. Variations must also be gradually accumulated to create the nerves that cause the heart to operate by beating, thereby pumping blood. The electrical signals must also be created that move across the nerves from the brain. Thus variations must accumulate to create the nerves needed for the heart. What is also needed is for the blood to be pumped, and variations must also create the blood, with its many components. Then variations must be accumulated to form the veins, arteries, and capillaries, each in the necessary location for it to perform its function. This process of developing the operations of each new creature must continue with variations accumulating for every new body part needed to form the new body that is a human and every new creature that exists.

There are massive problems that arise from this creation process of accumulating variations. A new creature cannot function until all the parts are fully created and operating. The "unit" that is the body, with its ten major systems, cannot operate until then. It can only decay while waiting to operate.

Variations and Accumulations All at Once?

In his 2003 book, *In the Blink of an Eye*, Andrew Parker, a Royal Society research fellow at Oxford University's Department of Zoology, tells us:

> Internal organizations are under the control of many more genes, which *all* have to mutate at the same time to initiate a new internal body plan...internal body plans cannot be build up gradually because usually they can't function in intermediate states...internal body plans cannot be constructed step wise, and so are less influenced by the environment. Hence, convergence of internal body plans does not occur. [Italics in original][398]

Parker does not define "internal organizations," but by any definition it includes the ten major body systems, which include the systems for standing and moving (skeletal, muscular); the systems for energy and waste disposal (digestive, respiratory, circulatory, and urinary); the systems for coordination and control (nervous, sensory, and endocrine); and the systems for producing new life (male and female reproductive systems). All these systems must be created "at the same time to initiate a new body plan." These systems practically include the entire body. This is as close to admitting instantaneous creation, called saltation, as it comes without using the actual words. If you count the systems that need to be created, it comes to eleven when you consider the male and female reproduction systems separately. There are many other systems that must be created, but these are the major ones. For example, the visual system has many hundreds of parts to it, but it cannot operate or survive without the major systems. Another example is the teeth, which also cannot operate or survive without the major systems. The body is one unit, one system of systems. Not one part, organ, or system existed when the single cells were created. According to Darwin, they all were formed by variations (mutations) accumulating gradually. Thus, every creature is composed entirely of mutations, including humans.

A creation model does not operate differently on the inside of a body than outside of it unless that model shows how the cause and effect operate differently. Natural selection does not contain any means of operating differently due to body location. It does not contain a "location identification" capability. Random variations are created, and the process continues accumulations, which is all that is shown in natural selection. It really does not matter if a system that is being created is internal or external, for purposeless mutations must still be created, positioned blindly in a body location, aligned, accumulated, and be integrated with the other variations and body parts before it can possibly function, if it ever functions at all. In addition, the environment does not "influence" genes to mutate in a "direction" that is to become a specific series of variations that accumulate in a specific "direction."

Variations: No Anchor in Nature's Physical Reality

Due to the chance nature of variations' creation, it is not only possible but likely that all variations will be forever harmful for the creation of any organ, even in theory. This fact eliminates all gradual creation models from ever being science. The fact that the terms "harmful" and "mutation" have no dependency on nature, it removes any anchor to the physical world of nature. Mutations lack credible definitions and are manipulated by testimony in ways that are so removed from nature's reality that creation becomes surrealistic. Pierre-Paul Grassé is a renowned French scientist and past president of the Academie des Sciences. He is the editor of the thirty-five-volume *Traite de Zoologie* published by Masson in Paris. During his long life as a zoologist and biologist, Grassé has written several books and many papers in his chosen field.[399] In William J. Bauer's review of Grassé's book, Bauer finds that Grassé's purpose in writing his book is concisely stated in Grassé's own words. It is also a statement about natural selection:

> Today our duty is to destroy the myth of evolution [by natural selection], considered as a simple, understood, and explained phenomenon which keeps rapidly unfolding before us. Biologists must be encouraged to think about the weaknesses and extrapolations that theoreticians put forward or lay down as established truths. The deceit is sometimes unconscious, but not always, since some people, owing to their sectarianism, purposely overlook reality and refuse to acknowledge the inadequacies and falsity of their beliefs.[400]

Grassé echoes this observation and writes:

> The opportune appearance of mutations permitting animals and plants to meet their needs seems hard to believe. Yet the Darwinian theory is even more demanding: a single plant, a single animal would require thousands and thousands of lucky, appropriate events. Thus, miracles would become the rule: events with an infinitesimal probability could not fail to occur......There is no law against day dreaming, but science must not indulge in it.[401]

Lacking any anchor in physical reality, the task of the supporters of natural selection is to invent arguments, debate opposition, and, through interpretation, make natural selection appear credible. However, the role of natural selection is like that of a statue, an idol, attributed with characteristics that are not defined and do not exist in the model, nature, or science.

Variations: Adding Nothing to Our Knowledge of Creation or Evolution

Blankly staring out of Darwin's book is the fact that the model of natural selection does not tell us anything about creation or the operative parts of nature that are doing the creating. Darwin's writings about favorable or good variations and injurious or bad variations reveal nothing about nature or about what is supposed to have taken place. In his *Origin of Species*, Darwin shows us how little is known about the variations he uses in natural selection. He starts by writing that food causes creatures to change. But no one has ever demonstrated how arms, legs, nerves, or any body parts are caused by different foods, as if that were possible or ever observable. Darwin writes:

> There is also some probability in the view...that variability may be partly connected with excess of food.[402]

He then writes that variations and the processes surrounding them are not understood or defined:

> The results of the unknown, or but dimly understood, laws of variation are infinitely complex and diversified.[403]

In addition to not knowing what a variation is, Darwin acknowledges that the cause of each variation is not known:

> We are profoundly ignorant of the cause of each slight variation or individual difference.[404]

> We are far too ignorant to speculate on the relative importance of the several known and unknown causes of Variation.[405]

Throughout the six editions of *Origin of Species*, we are shown little to nothing about variations, their creation, how they accumulate, what selects them, how they are directed, or even what direction means. This is true to this day. Creation by natural selection is purely a testimonial exercise that is removed from science. Darwin was not troubled by the little that was known about variations or evolution using them. But he was troubled about the little that was known about creation in nature when special creation is accepted. He ignored the fact that special creation is a religious exercise and natural selection is supposed to be science that was observable in nature. Darwin writes that nothing is added to our knowledge when citing the Creator as the cause of creation:

> But many naturalists think that something more is meant by the Natural System; they believe that it reveals the plan of the Creator; but unless it be specified whether order in time or space, or what else is meant by the plan of the Creator, it seems to me that nothing is thus added to our knowledge.[406]

The same "nothing is thus added to our knowledge" that Darwin said of special creation is true for all testimonial models, including natural selection, as it contains no definitions, no parts of nature, no rules, and no relationships, which are everything that a science model possesses. A person who uses natural selection must imagine what is taking place and use inferences or other rhetorical approaches. However, inferences are not causal; they do not cause anything to happen outside of the imagination.

No Residual Trail of Variations or Mutations

When natural selection is believed to be the creation model, it follows that your entire body is created from mutations. Mutations are the only type of variation that is theoretically capable of providing the "raw material" for creation of new creatures. This poses a problem for supporters of natural selection. If you look at your skin, fingers, hands, feet, and face in the mirror, you will see no variations or residual marks from accumulated variations. If variations were accumulated to form body parts, there should be some residual marks on various parts of the body. Yet, we see no residual trail of variations to show they ever existed. Such assertions about seamless accumulations may be theories, but they are religious theories, not theories of science. It does not help that mutations are not defined in the model of natural selection or in nature. The term "mutation" is independent of nature, making it open-ended and dependent on a person's testimony. There is nothing on a fossil, human, or animal body to show variations ever existed, for there are no residual marks of any kind. Nothing in nature identifies what characteristics make a good or bad variation. There is no defined process to show how nature accumulates good variations in a "direction" toward a new body part for a specific creature. There is nothing to show that good or bad variations or their accumulations ever existed.

Variations: Physical Characteristics

Due to being randomly created, mutations are not necessarily homogeneous in their composition. The physical characteristics of the variations and the components within the variations include materials, shapes, dimensions, locations, orientations, strengths, flexibility, hardness, components, and so on. For example, a variation may contain the parts of a stomach, liver, kidney, heart, nerves, veins, glands, blood vessels, and other body parts. Each variation is said to be attached to an "accumulation," and the accumulation is attached to a newly forming part of the body. To have two variations successfully accumulate requires that each component of one variation exactly match each component of the next variation: nerves must align exactly with other nerves, organ variations must precisely align exactly with the corresponding organ accumulations, and zonules for the eye's lens case must be precisely created and located to move to the eye's completion. If a mismatch of one variation with another occurs, the accumulation fails. The creation process must start over again from the beginning. As an example, a variation that formed part of the chest cannot successfully accumulate with another variation that formed part of the feet, yet that is likely to happen as variations (and their locations) are randomly created. In another example, a variation that forms part of the eye cannot be used to form part of the heart, but that is as likely to take place as nothing in natural selection prevents it. As each variation and each component within the variation is created by random chance, there exists an infinite number of ways that each variation and each of its components may be created, including the location in the body. Each randomly created variation is uniquely different from every other one. The chance of one variation successfully matching and aligning with another variation or accumulation is zero. It is zero for every variation and every attempts at accumulation.

Rules of Variation's Accumulations

Many rules exist in nature's processes involving the relationships that cause the creation of each of the new creature's effects that we observe. One rule may require that simple arithmetic take place in a particular process, such as one volume plus one volume equals two volumes. Other rules may require that one volume plus one volume equals less than two volumes or more than two volumes. We see these types of rules operate in chemistry and physics. Each model of cause and effect contains rules by which the model shows how nature operates: rules exist for making chemicals, materials, shapes, and architectures and designating bodily locations, but natural selection shows no such rules and thus cannot be used in the creation of new body parts or new creatures, thus invalidating its being a creation model in nature.

The same is true for body systems creations. The organs in any system must be arranged in a specific sequence, which is based on the functions it performs. Every organ in the body performs its task after the one before it and must complement it if the body is to successfully function. The digestive system requires chewing first, swallowing next, and then muscular operations that allow the food to pass down to the stomach and beyond. Sphincter muscles act as gates, opening and closing to control the flow of the food and prevent it from moving backward. Each part of digestion is processed by a specific organ function, such as coating the stomach and generating acid for food digestion. When we look at natural selection, there is nothing in it that shows sequences of operation for any system in the body. Nothing is shown for how the many systems of the body operate together or their sequences. The rules of nature are missing for how the sequences of operations are created and the functions perform. In fact, all rules of nature are missing from natural selection.

In place of the missing rules of creation in nature, there is case-by-case ad hoc testimony to fill in for all that is missing in natural selection: that testimony may at best convince others about "how to think" about nature. Using this approach for "how to think" about creation and nature, the only thing that is achieved is the elimination of science.

Is Gradual Accumulation of Variations Even Possible?

It was the father of vertebrate paleontology, the renowned French comparative anatomist George Cuvier (1769–1832), who pointed out that creatures were whole units. Each creature is made of organs that are dependent on each other and work together. Each organ is part of one system. Each system is part of a system of systems. The system of systems works as one unit in the body, all the way down to each cell. Because they work together as one system, removing any one organ or system stops the others from operating to support the creature's body. Said differently, a gradual mutation of organs stops the system of systems (the body) from operating: death follows, not new creatures. Each organ contributes to the processes of the system it belongs to in a fixed sequence of operations. Each organ and system produces and exchanges products that enabled the body to continue operating in a fixed arrangement. The end result is that all the systems function together *as one unit* for the purpose of performing the totality of functions called "life." If one or more functions are stopped by the gradual or sudden changing of one or more organs or systems, then life ends for that creature: no evolution takes place.

There are ten major systems in the human body, differing for males and females. Some of these major systems are skeletal, circulatory, digestion, male reproductive, female reproductive, respiratory, nervous, and endocrine, among others. Each of the ten major bodily systems works with the others, such as the digestive system transporting food and nourishment to the circulatory system, which transports it to the rest of the body. The respiratory system contributes oxygen to the circulatory system that is transported throughout the body while the wastes are removed, including carbon dioxide, which is brought to the lungs and exhaled. If one of these systems were to be removed or prevented from making its contributions to the others, the body would perish. This failure of one function to be performed would take place during the time one or more organs was being changed to a new organ, to be part of a new creature that was being created. The body would perish, and all of the systems and organs would stop operating and perish as well when any changes were made. Because of the organs' dependency on each other and their working together, the body is shown to be a single unit that is one "system of interoperating interdependent systems," or simply a "system of systems" that is one unit.

Knowing about this unity of systems' support for each other, we see that the gradual creation of the unit is impossible. The probability of change to this biological "unit", as in the creation of a new creature, is zero. Changes must be instantaneous—*all* at once to all the components of the system: all the parts of the "unit" must be changed simultaneously, as in saltation. This is not an endorsement of saltation, but a denial of any gradual changes forming a new creature. For example, the digestive system could not survive apart from the circulatory system, and no system could survive without the nervous system connected to all the parts of the body. The male and female reproductive systems would fail without each other being fully operational, and both would fail without the other major systems in the "unit". So, too, would the gradual changing or creation of new organs from the "old" ones fail in any attempt to form a new creature. One gradual change of a system would destroy all the processes of the body's unity. Darwin never addressed this challenge involved with gradual "creation." But he did note that gradually created creatures were never found amongst the fossils. This was known in Cuvier's time. In 1827 Cuvier writes:

> Every organised being forms a whole, a peculiar system of its own, the parts of which mutually correspond, and concur in producing the same definitive action, by a reciprocal reaction. None of the separates can change in form, without the others also changing; and consequently, each of them, taken separately, indicates and ascertains all the others.

Thus, if the intestines of an animal are so organised as to be fitted for the digestion of flesh only, and that flesh recent, it is necessary that its jaws be so constructed as to fit them for devouring live prey; its claws for seizing and tearing it; its teeth for cutting and dividing it; the whole system of its organs of motion, for pursuing and overtaking it; and its organs of sense for discovering it at a distance. It is even requisite that nature have placed in its brain the instinct necessary for teaching it to conceal itself, and to lay snares for its victims.[407]

Cuvier defined an interdependent interoperating system of systems with each system having interdependent interoperating groups of parts. Gradual changes to bodily parts and systems would be disrupted and fatal to the creature being changed gradually as well as to the one being created. Because of the impossibility of gradual modification being used to form new organs from previously existing ones, Darwin should have addressed this obstacle, which is a fatal flaw in his model of natural selection. Unfortunately, Darwin never addressed this requirement for unity of operations except in one case. He writes:

If it could be demonstrated that any complex organ existed, which could not possibly have been formed by numerous, successive, slight modifications, my theory would absolutely break down.[408]

That "complex organ" which could not possibly have been formed by numerous, successive, slight modifications" is the human body—and all other creatures' bodies as well. This "unity" is one more reason that gradual creation is impossible. This impossibility is the reason that the incomplete intermediates are missing. It is the reason that only fully completed creatures are shown as intermediates.

Cuvier demonstrated a unity of the body prior to Darwin's publication of the 1859 *Origin of Species*. It is repeated here with more detail, showing the reasons for the failure of Darwin's theory. This unity of body parts and systems that function as one single system is as true for one cell as it is for the each different creature's body, such as for a fish, reptile, bird, whale, or human. Even in theory, the changing of a wolf like creature into a whale is impossible with gradual creation, yet Darwin thought it was possible. Today, natural selection supporters believe it actually took place. To admit otherwise would be to admit natural selection is a failure—as if their acknowledging that fact is necessary to draw that conclusion. The combined operations of the body's systems provide for the very different process of sustainment (food distribution, waste packaging, and waste removal), maintenance (repair and replacement of parts), operations (nerves allowing actions), and reproduction (which necessitates that all the other systems be operational). Changing one organ or system is likely fatal to the body and to Darwin's model of natural selection—his "theory."

Consider the creation of the male and female reproductive systems that were created separately in two different bodies, yet must operate together as one system. Each has very different parts that perform many different functions, yet the two function as one system to enable the creation and nourishment of offspring. If any attempt to change the system in the two bodies were made to the ovaries, fallopian tubes, uterus, sperm, or any of the related parts, reproduction would end long before any new accumulation of different variations could be used to create a new reproductive system. No gradual change of bodies is possible to any system that is part of the "unit" that is the body. This is true for each creature's male and female reproductive systems. No reproduction could take place during the many millennia when gradual change was taking place. The end result is changes to the old reproductive system in the ancestral creatures ceases to function because they have been changed and the new reproductive system never reaches a point where it is created and functioning. This would be true of every system, not just reproduction.

This unity of systems was known in Darwin's day, as it is today: integrated interoperating systems cannot be changed gradually, regardless of the cause of creation. Neither Darwin nor Wallace addressed the unity of systems operations and how natural selection was impacted by changes to the organs or systems. Nature's cause and effect model of change by which critical organs and systems gradually change has not been addressed to this day. Perhaps it is not addressed because to discuss it would be to show that gradual creation is impossible. Indeed, the body cannot have its organs changed gradually to become another creature, yet this is exactly what was claimed by Darwin and is believed to this day by supporters. The impossibility of changing a body that is a unit is taught in the biology and science textbooks as taking place under the umbrella of gradual change.

Variations' Identification

If variations existed as Darwin claimed, they would be physically identified rather than as they now are—imagined as logical entities that have no existence in nature. They would have been revealed in the fossils by all new creatures' organs being shown in incomplete, gradually accumulating stages, including the gradual accumulations of different skeletons, which appear fully formed and in their final architectural arrangement. One would be able to determine how many variations are needed to create a part of a body, such as a bone, nerve, electrical signal, or hormone. The single stubborn fact remains that fossils that are declared to be missing, for any reason, are simply non-existent; they are no different than fiction. The other single stubborn fact is that the fossils appear exactly as they should when creation by saltation is operating.

Claiming that the missing variations and the partially formed intermediates once existed is a declaration of beliefs and religious faith, not science. Darwin's missing fossils may as well be spirits for the lack of reality they show in nature. The missing incomplete fossils are actually spirits of the mind held as existing by the worldview and the faith that reality is wrong and the theory that contradicts reality is correct. In the end, Darwin's argument of a new unseen incomplete creature being found between two completed creatures is a thinly veiled religious, faith-based argument whereby design was marched in the back door by Darwin. Such incomplete creatures have never been shown to exist, despite the vast numbers that should be observed.

181

Normal Distribution of Variations

There is no astounding revelation in the fact that all body characteristics vary in measurement, for it is apparent to everyone who looks, especially parents. The variations that exist between a parent and child oscillate about a common mean; they do not take off in a random or "selected" direction. For example, bodily characteristics that oscillate about a mean are height, weight, hat size, shoe size, waist size, inseam length, chest size, hip size, nose size, organ sizes, fixed architectural arrangement, intelligence, and so on. These variations never create new body parts, and they are passed from grandparents to parents to offspring and their genetic recombinations. Mutations are different from recombinations of genes in that recombinations are mixtures of *existing genes* from each parent, but mutations are the accidental creation of *new genes* that did not previously exist before their creation. Accumulations of mutations become new body parts by selection and direction, or so it is believed. This is the reason mutations cannot possibly form a normal distribution: the bodies are incomplete for thousands of years, and the mutations do not oscillate about a mean but rather are said to form in a direction. That direction cannot be "normally distributed." With mutations accumulating into new organs, there is no average or standard deviation from that average. There is no population from which a normal distribution could be representative, even in theory.

Normal distributions show only one aspect of evolution: *stasis about a common body shape.* They show that no evolution is taking place, for no mutations are being introduced into the distribution. For example, an average height can be computed from all the individual heights of people in a large group. When the difference between the average height and each individual height is plotted on a chart, a bell-shaped curve is drawn. This bell-shaped curve is called a "normal distribution." Heights may vary, but no amount of variation in height can create a new creature. No amount of variation in beak size can create a new body part or new creature from a finch; no evolution takes place at all with finch break variations, which are a celebration in stasis.

THE MEANING OF VARIATIONS

Even if one found height varying in one direction, nothing new is being shown. If nose sizes were growing in one direction, like beak sizes, they only vary within specific limits. The finch beak variations illustrated oscillating variations for both David Lack's study and the subsequent Grant's study as well: every part of every body varies between fixed limits, with no new body parts being created and no evolution taking place. One may find names given to normal distributions that may be misleading, as is the case with "directional selection," "stabilizing selection," or "disruptive selection," but they can never mean more than variation about a common body plan. It is misleading to label a normal distribution as "selection," implying that evolution is taking place. There is no new genetic material being created that would permit evolution to take place, even in theory. Not only is such a practice technically incorrect with regard to evolution, it is technically incorrect with regard to normal distributions.

Summary

There are 1,014 different types of cells making up over two hundred different kinds of tissues[409] in the human body. Each of the 1,014 different kinds of cells in the human body—in the brain, liver, bone, heart and many other structures—are reading off a different set of the hereditary instructions written into the DNA.[410] No DNA existed on earth for about 1.2 billion years after the earth's beginning. Then came the sudden appearance of single-celled creatures with their genetic code. In the Darwinian worldview, mutations of DNA were the cause of each new creature that appeared afterward. Such new creatures involved countless, if not unimaginably large, number of undefined mutations being created at random. Each of the new DNA components and arrangements that formed the new, living single-celled creatures never existed prior to their first creation. Each of the new body parts never existed until mutations accumulated and formed them by an undefined process and in an undefined "direction." Each of the cell architectures was mutated into existence, according to testimonies that are given. Yet, natural selection does not contain cells, tissues, or anything needed to create them. It does not contain the rules or relationships that could create the cells and functions that are performed in or out of those cells and tissues. Consequently, natural selection could not have created them, and only the faith-based attributions link their creation to natural selection.

With natural selection, the creation of a bodily organ could only take place gradually by mutation in the first part of natural selection. Take the heart, for example; it could only be formed by heart variations being accumulated with other matching heart variations that have formed an accumulation, which is a partial heart; similarly, a lung could only be created by lung variations accumulating with other matching lung variations forming an accumulation, which is a partial lung; and, each organ could only be accumulated with like-variations for that organ, which is a partial organ. Yet, we are only given metaphors called "selection" and "direction," which are mere figures of speech as parts of a person's testimony. There is no reality to these metaphors of evolution; yet, they are supposed to be accepted as science, as the way creation takes place in nature. Is this fiction being testified as being fact?

Without exception, every causal model of evolution, before and after Darwin, used the testimonial approach to render the appearance of operating in nature. The field of evolutionary biology uses testimony exclusively in their creation models (which are said to be responsible for evolution), including inferences about the past. Even the claimed biological relationships between fossils are testimonial as natural selection tells us nothing about biological relationships, time, kinship, or appearances—all inferences, all non-causal.

The approach of using testimony to support a model of creation is an approach used by all major religions, including the religion of natural selection, which its supporters call "Darwinism." It is testimony that links cause and effect in religious models: testimonial cause to physical effect. The religion called Darwinism has natural selection as its core tenet, and it is faithfully held to be responsible for creation. Atheist Michael Ruse was a key witness in the 1981–1982 Arkansas "balanced treatment" case, and he writes about his court experience and how evolution by natural selection is a religion:

I still remember arguing in the Arkansas court house with one of the most prominent of the literalists (now generally known as creationists). Duane T. Gish, author of the best-selling work, "Evolution: The Fossils Say No!," resented bitterly what he felt was an unwarranted smug superiority assumed by us from the side of science.

"Dr. Ruse," Mr. Gish said, "the trouble with you evolutionists is that you just don't play fair. You want to stop us religious people from teaching our views in schools. But you evolutionists are just as religious in your way. Christianity tells us where we came from, where we're going, and what we should do on the way. I defy you to show any difference with evolution. It tells you where you came from, where you are going, and what you should do on the way. You evolutionists have your God, and his name is Charles Darwin."

At the time I rather pooh-poohed what Mr. Gish said, but I found myself thinking about his words on the flight back home. And I have been thinking about them ever since. Indeed, they have guided much of my research for the past twenty years. Heretical though it may be to say this—and many of my scientist friends would be only too happy to chain me to the stake and to light the faggots piled around—I now think the creationists like Mr. Gish are absolutely right in their complaint.

Evolution [by natural selection] is promoted by its practitioners as more than mere science. Evolution is promulgated as an ideology, a secular religion—a full-fledged alternative to Christianity, with meaning and morality. I am an ardent evolutionist and an ex-Christian, but I must admit that in this one complaint— and Mr. Gish is but one of many to make it—the literalists are absolutely right. Evolution is a religion. This was true of evolution in the beginning, and it is true of evolution still today.[411]

Ruse clearly determined that evolution by natural selection is an ideology, a secular religion. That point has been made here repeatedly. It could not be otherwise because natural selection has none of the characteristics of science or nature: no physical components, processes, rules, relationships, or nature's links of cause to effect. Yet, evolution by natural selection is presented in biology and science textbooks as if it were science. In those textbooks, natural selection is part of a religion that is dressed as "science"—the only religion to do so. It is a "science" that does not operate or exist in nature.

A religion has a number of identifying characteristics that differentiate it from science. All religious models deal with man's origin, purpose, morality, and destiny. Theories that do not deal with these four topics are not religious. The first observable characteristic of a religious model is that it has no physical components in it, as is the case with natural selection, miracles, and special creation. Actually, a creation model's inclusion of only one component that is independent of nature relegates it to being a religious model. A second identifying characteristic of a religious model is the link between cause and effect being testimonial, such as the link between natural selection (cause) and an adaptation (effect); or between God (cause) curing a terminally ill person being healed (effect). That link cannot be found in nature as a physical part of the creation model. The link may be established by testimony using causeless rhetorical processes, such as by extrapolations, correlations, inferences, and other rhetorical means. The absence of a foretelling capability in both natural selection and miracles necessitates that the testifier *attribute* his or her personal prediction or observation for what is to take place to natural selection (in the case of evolution). A third characteristic of a religious model such as natural selection is that it cannot foretell what is to take place. That is, it cannot foretell what is to be observed and then confirm it by observation. All models of science foretell what is to take place. A fourth characteristic of religious models is that the terms in the models are independent of nature.

That is, the terms are open-ended and change to suit each situation. Examples of such terms are "competition," "nature," "adaptation," "fitness," "chance," "variation," "selection," and "direction."

The models that are called "science" have rightly achieved their status because they are used to create technology that is deemed desirable by people, such as vaccinations, phones, televisions, radios, automobiles, and aircraft. Our lives are changed by the application of science to nature to create technology. In contrast, no testimonial model has ever been responsible for the creation of any technology or anything useful. Testimonial models are incapable of such applications. That is the fifth characteristic of a religion—not being capable of being used to create useful things in the physical world of nature. The sixth characteristic of religion is that morality comes from absolute sources (i.e., Bible, God) or from feelings, which is a relative source. Science models cannot determine morality, whether absolute or relative.

The debate on the causes of evolution often includes claims that science models are more important than testimonial models. It may also take the form of science being more important than religion. This is not the case. It is impossible for life to be lived by the models of science because values, morals, feelings, and tastes are independent of nature—they do not exist in nature but rather in people. All lives are lived by testimonial models, not the models of science. In addition, testimonial models are more important than science models because they change what is inside a person. They change how a person lives, views life, and relates to others. To illustrate the importance of testimonial models, science was used to create the atomic bomb, but testimonial models were used to decide to drop it—and on whom. The value of saving a net amount of lives was the justification. If Hitler or Imperial Japan had been first with the atomic bomb, other catastrophic results would have been obtained because of the evolutionary morality that would have been used by those nations. Hitler's gas chambers would have been the first step toward an unimaginable horror.

In another example of testimonial models being more important, Hitler used morality from "survival of the fittest" which he used for a "way to think" about people, murdering many millions in accordance with that "way to think." The same may be said about Joseph Stalin and many others who were responsible for murdering more people than Hitler. In another example of testimonial models being more important, Mother Theresa had a testimonial "way to think" about people that used an absolute morality from the New Testament. Hers was a life well lived. There is no science to morality, and none can ever be invented, because morality does not have the characteristics of science.

CHAPTER 10

SELECTION PRESSURE, ARMS RACES, AND OTHER INCANTATIONS

Introduction

Evolutionary biology models do not contain biology or biological processes. They do not operate in nature's physical world, and they do not show biological cause and effect—they only show "testimony and effect." They are portrayed as models of how creation that forms evolution takes place, causing bodily changes such as new limbs, eyes, mouths, claws, nerves, electrical signals, chemicals and chemical reactions, and the arrangement of body parts and the body's major systems. The evolutionary biology models presented in this chapter reflect many, if not all, evolutionary biology models: each has a fatal flaw. Each model is a *metaphor*, a tool of rhetoric, a manner of expression, but not a part of nature that operates in a cause and effect manner. The models are *incantations* that are spoken in place of showing that nothing is known about what is taking place in nature. The following examples of evolutionary models that are merely incantations include selection pressure, selection forces, arms races, adaptive landscapes, and personification. With natural selection, these models form part of evolutionary biology—without a hint of biology in any of them.

1. The Selection Pressure and Selection Forces Incantations

In 1872, one of the first exploratory vessels, the HMS *Challenger*, traveled to explore great ocean depths that existed under enormous levels of pressure. The ocean's physical pressure and physical forces are related. Pressure

189

is a uniformly applied force per unit area, such as pounds per square inch. Forces that are not uniformly applied may be applied at one or more points, like one or more columns that apply forces at their bases. If a force is involved with pressure, then it must be spread uniformly over an area. During its four-year journey, the *Challenger's* voyages circumnavigated the globe, sounded the ocean bottom to a depth of 26,850 feet, found many new species, and provided collections for scores of biologists.[412] The *Challenger's* research findings started revolutions in earth science and biology over the next hundred years.[413] The great pressure in the depths of the ocean represented one lesson that surfaced from the *Challenger's* findings. Darwin's voyage on the *Beagle* paled by comparison. The equipment (trawl with a beam made of wood) used to gather samples showed the great physical forces exerted on it by the ocean,[414] as described in Stony Brook University's (SUNY) 2013 description of "The HMS Challenger Voyage."

> Consider the difficulties of sampling: One trawl was put over the side at 9:00 a.m. in a bottom sounded at 1950 fathoms (11,700 feet or ca. 3,600 meters; about two miles deep). The trawl was hauled in at 5:00 p.m. The beam was broken through the middle, and otherwise strangely torn and crushed, by the combined action of the pressure to which it had been subjected, and the strain of pulling it up rapidly through three miles of water. The wood was driven in and compressed so as to reduce the diameter of the beam by half an inch, and the knots projected a quarter of an inch on all sides.[415]

The physical force of the ocean's pressure was so great that it squeezed the trawl's wood and reduced its dimensions. As ocean pressure increases with depth, the ocean's increased pressure on an object becomes apparent: it becomes observably compressed, as the *Challenger* revealed. Although the ocean's pressure is measurable and pervasive in the oceans depths, another pressure used in evolutionary biology has no units by which it can be measured (such as pounds per square inch). Selection pressures have no measure, for they do not exist in nature and dwell only in an imaginary existence, never moving outside of the borders of the mind. The term "selection pressure" does not represent an actual pressure but rather a manner of speaking about nature. It is a "way to think" about creation. It is part of a growing lexicon of evolutionary terms that have no existence in nature's processes but are used to describe the Darwinian "way to talk" and a "way to think" about creation. In evolutionary discussions, as Ernst Mayr tells us, it is often stated that selection pressure has resulted in the success or elimination of certain characteristics, without telling how:

> Evolutionists[416] have used terminology from the physical sciences[417]…It must be remembered that the use of words such as force or pressure is strictly metaphorical, and that there is no such force or pressure connected with selection, as there is in discussions in the physical sciences.[418]

As Mayr states, selection pressures are metaphorical, not existing outside of a manner of expression. In nature, physical pressures cause physical forces to exist. Inversely, physical forces can cause physical pressures to exist. This is not the case with selection pressure and selection forces. A selection pressure is merely invoked and attributed to observations, but it does not exist in any actual physical sense in nature. Selection pressure cannot be seen to cause any events, and it is not defined in any operative way that can then be used to show that it creates new body parts in nature. The same is true for selection forces, for they cannot be shown to exist in a physical cause and effect model. Andrew Parker defines selection pressures as "a way of thinking," consistent with its metaphorical nature. He writes:

> [A] modification in the environment can be *thought of* as a pressure on the local animals to change. Hence the term "selection pressure" was introduced. [Italics added][419]

191

Varying degrees of selection are imagined to exist, with small selection pressures causing small changes and larger selection pressures causing larger changes in animals. These imagined selection pressures and changes are supposed to take place despite having no means of defining, identifying, or taking measurements as happens with pressures in the ocean's depths or elevations at mountain peaks. The products of selection pressures are always identified afterward by fiat: simple testimonial declarations. Such empty of nature claims are simple, but false. Parker apparently believes them when he writes:

> A minor selection pressure may result in a slight modification in local animals. An animal walking on the sea floor may develop slightly broader feet to prevent it from sinking if the sand and mud becomes softer. A weighty selection pressure may result in considerable modification in a local animal.[420]

Despite being a metaphor (nothing more than a mere thought or "way of thinking"), selection pressure is made to appear as a very real presence in nature, but it is not. The first eyes on earth were created over 500 million years ago. While parts of nature are not identified in any causal model of creation, fiat testimony is used for creation, as shown next:

> Precambrian selection pressures had been acting on proto-trilobites to evolve an eye.[421]

Selection pressure can be used to argue one's imagined views. In Jonathan Weiner's 1995 book, *Beak of the Finch*, he writes that selection—as in natural selection or selection pressure—is an argument, perhaps not realizing that arguments are not causal in nature; they merely serve as a tool for expressing one's beliefs and faith when it involves creation that leads to evolution. In his words:

> If you see two species that are different, you are always able to fabricate an argument that selection formed those differences.[422]

In this quote, we see the world of the imagination could easily be substituted for the physical world of nature: one merely makes arguments, evolutionary biology arguments. In the imagination, a pressure—even an admittedly undefined and nonexistent one—is said to cause changes to creatures. It sounds factual to the owner of the thoughts and perhaps to the listener or reader. The metaphor of selection pressure allows it to appear that walking on soft mud causes wider feet or that eating new foods causes changes in the structure of the mouth, jaw, and skull. The fact remains that natural selection does not contain mud, feet width, or new foods, which makes it impossible for it to even be considered to represent nature's processes or body part creations. Parker writes about selection pressure and how it creates new body parts, but he omits the operational details of nature's cause and effect. His claims remain a "way of thinking" and viewing the world of nature. Selection pressures become the new synonym for miracles.

For Parker, even sunlight is imagined to be a selection pressure that is responsible for creating eyes—but only for some creatures. Other stories must be created for those that do not have eyes: if some had eyes created by pressure, why not others? So too with soft mud: selection pressure causes wide feet for some, so why not others? Why any at all? The fact is that no physical relationship exists between soft mud and feet growing wider. No relationship is known between hard soil and narrow feet or between predators eating prey and the prey developing new body parts in response. If soft mud and sunlight are selection pressures, the scale for detecting the pressure never leaves the imagination, for they are not detectable in nature. Selection pressure is not measurable but is treated as "fact." In the physical world of nature, foretelling models show that pressure is very real. It causes events in a cause and effect manner. In the ocean's depths, pressure and forces exist, but no such force or pressure exists in evolutionary biology. Ernst Mayr explains:

> The metaphor of selection pressure is frequently used by evolutionists to indicate the severity of selection.[423] Even though it is a picturesque expression, this term, borrowed from physical science, could be misunderstood, for there is no force or

pressure connected with natural selection that corresponds to the use of the term in the physical sciences.[424]

In the physical world, we have rulers to measure distance, scales to measure weight, devices to measure volume, temperature scales to measure heat, electrical meters measure current and voltage, and light meters to measure brightness. In evolutionary biology, selection pressures, however, have no such units or devices for measurement, for one cannot measure what has not been shown to exist in nature. Without showing a cause and effect relationship in a model of creation, Parker claims the pressure of sunlight was selective enough to create new eyes. He writes about this creation event:

> The first eye must have evolved in response to an increase in *sunlight*, a factor independent of evolution[425]...In fact, animals with eyes may even provide the main *selection pressure* in the evolution of some plant groups. [Italics added][426]

Using selective pressure and selection forces, we know nothing about cause and effect in nature, for there are no such pressures or forces in nature. Neither selection pressures nor selection forces are related to science, nor is evolutionary biology as a discipline for the creation of new creatures.

2. The Arms Race Incantation

The metaphor of the arms race is a driving evolutionary force often cited today. Proponents portray it as ancient, as old as nature, taking place between a predator and its prey. It is a prominent selective force in the evolution of species[427] this arms race is said to have resulted in nine orders of trilobites[428] and an enormous number of families by the end of the Cambrian period.[429] Trilobites are an ancient, extinct ocean arthropod with flat, oval bodies and dorsal exoskeletons divided into three vertical sections.[430] In simple language, they have an external skeleton like a lobster or crab. Richard Dawkins writes about this driving evolutionary force as part of the arms race:

> Prey animals evolve faster running speeds because predators do. Consequently predators have to evolve even faster running speeds, and so on, in an escalating spiral. Such arms races probably account for the spectacularly advanced engineering of eyes, ears, brains, bat "radar" and all the other high-tech weaponry that animals display.[431]

As shown in this quote, the arms race metaphor becomes a "way to think" and a "way to talk" about creation and evolution, even though it is imaginary. Dawkins believes the actions of one creature alter the genetic material of another. It is belief that has nothing to do with science. With humans, engineering is usually performed by a person trained in many causal models of nature. In nature, however, supporters of natural selection believe that engineering is performed by one creature chasing after another, according to the arms race model; no intelligence is needed, just one creature chasing another. By this activity, the genetic material changes to create new eyes, ears, brains, bat sonar, and high-tech weaponry.

The actions of running, evading, seeing, and being seen thus engineer all the parts of each new body. Arms race incantations get credit for creating new genetic material with new biochemicals, nerves, and electrical signals, all positioned exactly where needed and operating as necessary for a creature to exist. The way arms races operate is not difficult to understand. When creatures proliferate, such as the trilobites, an imagined arms race becomes an imagined selection pressure and is given credit for observations of fossils. Selection pressures and arms races are a way of talking and thinking about fossils and living creatures, possibly meant to replace competing ways of thinking. An example using trilobites follows:

> With exceptions, including trilobites that lived in deep and dark places, the trilobites of the Cambrian already had a highly advanced visual system. In fact, so far as we can tell from the fossil record thus far discovered, trilobite sight was

far and away the most advanced in Kingdom Animalia at the base of the Cambrian, providing a decided survival benefit of being able to see both food, as well as other creatures for which they could become food.[432]

Arms races are no more than metaphors and only exist in nature through faith-based testimony. Arms races are imagined to create new genetic material that create new body parts, such as nerves, organs, eyes, claws, anatomy, and bodily materials. The absence of biology in the arms race model makes it impossible for the model to do anything in nature, yet it is credited with creature creations as mystically as miracles create new creatures. Alfred Russel Wallace, the co-discoverer of natural selection, used the arms race as a cause of new creatures in the same way it is used today: simply by declaring that it took place. In his 1889 book, *Darwinism,* Wallace describes the arms race without using the actual term:

> Climate too has changed again and again, either through the elevation of mountains in high latitudes leading to the accumulation of snow and ice, or by a change in the direction of winds and ocean currents produced by the subsidence or elevation of lands which connected continents and divided oceans…while the change from a damp to a dry climate may necessitate migration at certain periods to escape destruction by drought.[433]

With these changes to nature (climate, elevation, snow, ice, wind, ocean currents, land mass changes), Wallace imagines that creatures are forced to change, as in today's claims for the arms race. Wallace writes:

> This will expose the species to new dangers, and require special modifications of structure to meet them. Greater swiftness, increased cunning, nocturnal habits, change of colour, or the power of climbing trees and living for a time on their

foliage or fruit, may be the means adopted by different species to bring themselves into harmony with the new conditions; and by the continued survival of those individuals, only, which varied sufficiently in the right direction, the necessary modifications of structure or of function would be brought about.[434]

Wallace imagined that the interaction between creatures creating greater swiftness, increased cunning, nocturnal habits, and change of colour, and so on, to bring themselves into harmony with their new conditions. For Wallace, creations were caused by climate, elevation, snow, ice, direction of winds, ocean currents, land, and competition. The creatures ran a race with nature, which, in Wallace's mind, required creatures to undergo "special modifications of structure."[435]

Without being able to cite genetic or biological material in a physical "cause and effect" relationship with nature in the model, one must change "how to think" about creation to find agreement with Wallace's claims: this means one must change his or her view of creation, which involves changing his or her belief system to Darwinism. Nowhere in nature does creation of new creatures take place because of something called "nature" or because of a creature's actions. Creation of new creatures mandates that genetic material change in a precise manner, which then creates new variations that, with countless other new variations, accumulate in a direction to form body parts in the precise locations of each creature's body. Only testimonial models have ever been used to make such claims, for they do not have the restrictions of reality. Because of the immense complexity that a model of nature would, by necessity, have to show how creation takes place, not one model will ever possess the characteristics of science models. Incantations are all that will ever exist.

3.   The Adaptive Landscapes Incantation

Adaptive landscapes are not actual landscapes or geographic terrains in nature, and they cannot be measured by any surveyor for their elevation, depth, or location. They are not represented by topographical maps with contour lines drawn by surveyors and civil engineers or by graphs or charts taken from measurements of nature's characteristics, such as height, weight, distance, time, or the numbers of creatures living or dying. The metaphoric incantation of adaptive landscapes operates in the imagination yet is claimed to function as if the landscapes actually existed in the physical world of nature. Like all the other evolutionary biology incantations, it has no legitimate place in any textbook of science or biology. Adaptive landscapes provide ways of mentally portraying "how one thinks" about creation. Niles Eldredge and Ian Tattersall write about this metaphor:

> In 1932, the geneticist Sewall Wright developed a simple yet effective imagery to convey some of his ideas about relative "success" of genes in breeding populations. Each gene is a locus, or place, on a chromosome. Each gene has one or more forms or "alleles." With thousands of loci, each with several alleles, one would expect some of the many possible combinations of these alleles to be more "harmonious" than others, as Wright put it. The more harmonious combinations are those that produce the *fitter* individuals—the ones better equipped to flourish in the environment.[436]

The terms used by Wright in the above quote—"success," "harmonious," "fitter," and "better equipped"—work independently of nature and do not operate in the physical world. That is, these words do not exist in the world of nature. They promote "ways of thinking" about nature and genetics and "ways of viewing" nature, but they cannot show how nature operates. One does not know any more about what happens to cause creation after using this incantation than before using it. By its very imaginative characteristics, Wright's metaphor of adaptive landscapes could only equate to rose-colored glasses to shape his personal view of how life is created. He would like others to adopt his view. The use of an imaginary "way to think" did not stop Wright from applying his metaphor as if actually existed in nature. Eldredge and Tattersall continue:

But Wright posed a problem and to dramatize *it* he drew a crude map. Hills and valleys were delineated by contour lines. On conventional topographic maps, the contour lines connect areas of equal [metaphoric] elevation. On Wright's concept of the "adaptive landscape," the contour lines connected hypothetical [metaphoric] regions of fitness. On the tops of the hills sat the more harmonious allelic combination: in the valleys reposed the less fit individuals. The problem, as Wright saw it, was this: how does a species maximize the number of individuals sitting on the peaks?[437]

The metaphoric terrain and terms Wright used no more exist in nature than could "high," "low," "deep," "shallow," "big," "small," "hot," "cold," "far," "near," "fitness," "selfish," or "altruistic." No model of nature can be created using these terms, but they are used in biology and science textbooks. The terms may compel some minds and imaginations, but they are completely unable to portray nature in any physical cause and effect role of creation. Eldredge and Tattersall acknowledge the effect that the metaphor may have on people:

> The power of this imagery is compelling. Others soon picked it up and used it for purposes far beyond its original intent. Today, no textbook on evolution is without this *pictorial metaphor*—and seldom is the difference between Wright's original use and the more typical notions developed in the 1940s pointed out.[438]

When this metaphor appears in a textbook of science, biology, or nature, it misleads students to thinking this is science or biology. It may possibly indoctrinate them or inculcate them into false beliefs and false "ways of thinking" by statements that have a foundation in mysticism and in worldviews that cannot be shown to operate in nature or science. This imaginary metaphor of landscape cannot even be considered a candidate for science as it does not exist or operate in nature. A metaphor, as a means of explaining ideas, can be useful if the actual operative model is known and exists in nature. This is not the case with adaptive landscapes or any other evolutionary biology incantations. Eldredge and Tattersall continue:

> The adaptive landscape in its familiar guise sees the hills and valleys as environments to which populations or species of organisms are adapted. The hills are ecological niches to some and simply peaks of adaptive perfection of the occupant population to others. The valleys are inhospitable areas unoccupied for any length of time: literally, valleys of the shadow of death. The difference between this image and Wright's original metaphorical geography is twofold: entire populations of species occupy the hillsides, and there are environmental differences between adjacent peaks. Each peak is a different ecological niche. The problem then becomes: how do species cross the valleys and climb the next peak?

> As it was developed in the 1940s, the metaphor of the adaptive landscape is used to explain how adaptation [creation] occurs via natural selection. The entire history of life falls out from this picture of changing landscapes followed by adapting species, and this explains why the adaptive landscape is the favorite graphical image of the synthetic theory of evolution.[439]

The term "adaptation" that this model seeks to explain does not exist in nature as it is independent of it. The "explaining" in the quote means adopting a "way to think" with the person agreeing. Any "way to think" or necessity for agreement eliminates any possibility for science. So one model, empty of nature (adaptive landscapes), is used to explain an adaptive effect that also has no existence in the physical world. Adaptation exists independent of nature and receives its ad hoc meanings by case after case of attributions. Adaptations include hands, feet, eyes, balance, bipedalism, grasping, and so on. One can find no shortage of examples of definitions of adaptations by attribution, but not one definition depends on nature in this model, such as is found with terms like inches, pounds, cubic feet, temperature, germs, and so on.

No physical reality exists in the entire discussion of adaptive landscapes, regions of fitness, ecological niches, contour lines, adaptive perfection, or fitness. What is portrayed to exist in the physical world of nature exists entirely in mental imagining exercises. It is supposed to be cause and effect and science, not "testify and effect," "imagine and effect," and beliefs. The model of natural selection has no hills, no valleys, no environments, no adaptations, no biology, and none of the characteristics of the type of knowledge that are called "science." Because of the emptiness of nature in adaptive landscape, only metaphors, incantations, and testimonies can be used to support them: empty of the real world of nature, adaptive landscapes create nothing but wild stirrings of the imagination.

4. The Personification Incantation

The type of person capable of creating new forms of life is not an ordinary human but rather a superperson—one with extraordinary powers, one commonly called a deity, a god. Personification, when combined with creation of new creatures, portrays a deity. Personification omits all parts of nature. It omits naming the blind components that Darwin says created the new variations that accumulated to new forms of life. Darwin hid behind his rhetoric. He left the parts of nature that do the creating unnamed; he had to, because no such parts of nature have ever been shown to exist. Darwin often used personification, as is done to this day. There are several ways to use personification. One of them may be to deify a part or process of nature, to make a god out of it,[440] or to give something the ability to take the actions of a god, such as in creating new forms of life.[441] In a personification used in the first edition of *Origin of Species*, Darwin deifies natural selection to rival God in Genesis. He writes:

> In living bodies, variation will cause the slight alterations, generation will multiply them almost infinitely, and natural

selection will pick out with *unerring skill* each improvement. [Italics added][442]

If the claim is merely a metaphor, a way of speaking, then he should have given the actual explanation first and then used metaphors. But he never gave an explanation in any of his writing of how variations were created, what they were, how nature "selected" them, and how "direction" was determined. Darwin found personification of natural selection acceptable and supportive of a model that he self-servingly called "scientific." Darwin was soundly criticized for the practice of using personification, especially since the word "selection" implies that nature operates as a superperson—a deity—during creation. The term "selection" was taken from the analogy of humans selecting animals for breeding. Darwin responds to criticisms of deifying nature in the fourth chapter of the 1862 third edition of *Origin of Species*:

> Several writers have misapprehended or objected to the term Natural Selection. Some have even imagined that natural selection induces variability, whereas it implies only the preservation of such variations as occur and are beneficial to the being under its conditions of life[443]...Others have objected that the term selection implies conscious choice in the animals which become modified; and it has even been urged that as plants have no volition, natural selection is not applicable to them![444]

Nature does not select—people do. Nature is portrayed using models that reveal an interaction between its named parts, operating by the rules shown in the causal models that represent nature in acts of creation. Darwin's analogy and direct claims further show that the model operates as if it were a consciousness, if not a deity. In his writings, Darwin does not remove the reasons for the criticisms; he merely tries to make personification appear normal in "science." The defense is hollow and empty, but he makes it nonetheless in the 1861 *Origin of Species*:

> In the literal sense of the word, no doubt, natural selection is a misnomer; but whoever objected to chemists speaking of the elective affinities of the various elements?—and yet an acid cannot strictly be said to elect the base with which it will in preference combine.[445]

> It has been said that I speak of natural selection as an active power or Deity; but who objects to an author speaking of the attraction of gravity as ruling the movements of the planets?...No one objects to agriculturists speaking of the potent effects of man's selection; and in this case the individual differences given by nature, which man for some object selects, must of necessity first occur...Everyone knows what is meant and is implied by such metaphorical expressions; and they are almost necessary for brevity.[446]

Unfortunately for him, everyone does know what is meant: his model was inoperative in nature and faith-based, and he had to write something, even if false. Darwin's defense of using personification rests in citing metaphors used with chemistry, gravity, and agriculture as just a manner of speaking. He neglected to state that causal models in chemistry, gravity, vaccination, and other subjects all have causal models that are predictive and show nature's cause and effect. Metaphors like natural selection are empty and causeless; they reveal nothing in nature's world, just as the metaphor of natural selection reveals nothing about creation or evolution. The many personifications used by Darwin reveal him telling others what the model is supposed to do in nature, although the model is inoperative, telling nothing about the physical world's processes. The personification of a model does not operate on its own. Darwin's telling others what is taking place actually invalidates it as part of nature's operations in creation, for the model is mandated to do the "telling" by its itemized physical components, operations, rules, relationships, foretelling, and confirming observations. He uses one of his classic personifications as follows:

It may be said that natural selection is daily and hourly scrutinizing, throughout the world, every variation, even the slightest; rejecting that which is bad, preserving and adding up all that is good; silently and insensibly working, whenever and wherever opportunity offers, at the improvement of each organic being in relation to its organic and inorganic conditions of life.[447]

Using this version of natural selection's capabilities, neither Darwin nor the reader could tell what was taking place in nature. One merely repeats the empty phrases as if they were more than a deification of a model that has no parts of nature in it. In Darwin's following personification, breeders performed selection using intelligence, purpose, values, goals, plans, facilities, memory, written records, and judgments in their breeding practices. Darwin does not write about how these human characteristics are performed using nature's physical processes. Darwin does not distinguish between the breeders and nature; he endows nature with human and even godlike capabilities:

Can the principle of selection, which we have seen is so potent in the hands of man, apply in nature? I think we shall see that it can act most effectually.[448]

She [nature] can act on every internal organ, on every shade of constitutional difference, on the whole machinery of life.[449]

Notice that Darwin is testifying about the model's operations while it is inoperative, serving as an idol and contributing nothing. The emptiness of natural selection in nature necessitates analogies, metaphors, incantations, and personifications, attributing whatever is necessary to the model to make it appear operative, as in the following when Darwin tells his readers how natural selection operates. He must do this, as the model does not operate in nature:

[I]f it vary however slightly in any manner profitable to itself, under the complex and sometimes varying conditions of life, will have a better chance of surviving, and thus be *naturally selected*.[450]

I think it can be shown that there is such an unerring power at work, or *Natural Selection* (the title of my Book), which selects exclusively for the good of each organic being.[451]

Nature acts uniformly and slowly during vast periods of time on the whole organisation, in any way which may be for each creature's own good.[452]

[S]he [nature] may, either directly, or more probably indirectly, through correlation, modify the reproductive system in the several descendants from any one species.[453]

[N]ature cares nothing for appearances.[454]

We behold the face of nature bright with gladness.[455]

Each of the above quotes is a personification of nature, in essence a deification, complete with its likes, dislikes, preferences, capabilities, rules, relationships, and imagined operations. Seeing how he must attribute all the deified characteristics to the model, it is understandable when Darwin defensively writes:

That many and grave objections may be advanced against the theory of descent with modification through natural selection, I do not deny.[456]

Darwin could not show natural selection operating in nature—just the opposite of the scientific models of Newton, Pasteur, Mendel, and many others. Personification and metaphors are all Darwin had, which is the reason that the criticisms against Darwin and natural selection were warranted and legitimate. Realizing that nature is incapable of *selecting*, which requires values, objectives, decisions, standards, and other human characteristics, Darwin switched terms and used "survival of the fittest" in his 1869 (fifth) and 1872 (sixth) editions of *Origin of Species*. By switching names for his creation model, he substituted one metaphor for another, inviting more criticisms that last to this day.

Summary of Selection Pressure and Other Incantations

Evolutionary biology consists of creation models that are empty of nature, having nothing physical in them. They are merely invoked as if these models were the cause of creation, of adaptation, of new organs, of new limbs. Darwin defended his model of natural selection throughout his life. He had to argue and respond to arguments to give natural selection the appearance of legitimacy. He had help from others in that defense, most notably, in the beginning, from Lyell, Thomas Huxley, and the nine members of the infamous atheistic-focused X-Club started by Huxley.

Darwin updated his arguments in each edition of *Origin of Species* through its six editions, the first published in 1859 and the last in 1872. In the third edition, Darwin writes about his use of personification of natural selection and his use of metaphors as if they actually existed in nature. The flood of changes in the six editions of *Origin of Species* was a constant attempt by Darwin to respond to his critics, which is characteristic of testimonial models. The changes proved substantial. Morse Peckham recorded the changes to the *Origin of Species* as follows:

> Of the 3,878 sentences in the first edition, nearly 3,000, about 75 percent, were rewritten from one to five times each. Over 1,500 sentences were added, and of the original sentences plus these, nearly 325 were dropped. Of the original and added sentences there are nearly 7,500 variants of all kinds. In terms of net added sentences, the sixth edition is nearly a third as

long as the first. Of the total revisions, 7 percent appeared in the second edition, 14 percent in the third, 21 percent in the fourth, 29 percent in the fifth and the sixth had even more."[457]

It would be difficult to overstate the number of Darwin's changes through the six editions of the *Origin of Species*. Darwin's "science" changed often throughout his publications, even though nature did not. The editions necessitated a massive number of changes because *Origin of Species* is no more than a series of arguments that seek to change "how people think" and what they believe. Darwin's books never show how nature operates in a causal creation model. No model of creation is shown to take place in nature, yet that is what Darwin claimed; this empty of nature modeling is characteristic of all evolutionary biology models. The empty of nature terms of chance, good variations, bad variations, selection, direction, accumulations, selection pressure, arms race, and adaptive landscape merely encourage a "way of thinking" about nature. They become the empty phrases by those subscribing to the evolutionary biology worldview—a view that is not science. These terms are rhetorical tools used to give the appearance of something concrete in nature when nothing exists or operates as claimed.

The arguments used by Darwin necessitate the use of rhetorical approaches to appear legitimate, but science excludes such approaches. Legitimacy in science is shown by the model's predictions being followed by independent confirming observations. Historical inferences, which are part of incantations such as those used by shamans, are not causal or part of science, but they are the foundation of the religion of the supporters of natural selection. Natural selection, like evolutionary biology, depends on rhetoric, which is the art of harnessing reason, emotions, and authority through language, with a view to persuade an audience and, by persuading, to convince the audience to act, to pass judgment, or to identify with given values.[458] Historical inferencing is rhetoric meant to sway a person to the Darwinist "way of thinking" about nature. When people change their "way of thinking," they change their religion, even if unknowingly, but they are no closer to science.

According to Plato, rhetoric is the "art of enchanting the soul."[459] Natural selection is a rhetorical tool meant to enchant the mind, to have a person change "how to think" and "what to believe" about creation; it is meant to have a person possibly change his or her faith. Natural selection has the appearance of being a tool for proselytization, with the converted knowing no more about creation than before they were enchanted into new ways of thinking and believing. These new believers have to use *incantations* in place of nature, biology, science, and their former religions.

CHAPTER 11

TESTING MIRACLES AND NATURAL SELECTION

Introduction

Alfred Russel Wallace tested natural selection and, in so doing, served to illustrate the self-contradiction of the "scientific method" in evolutionary biology. He found natural selection incapable of the creation of some human body parts, such as the brain, hands and feet, skin, voice, and mental facilities. The testimonial tests he used for natural selection are no different in kind than those used for miracles and special creations, which is the reason for miracles being discussed first in this chapter. Wallace's tests of natural selection then follow.

Testing Miracles

An event that is called a miracle is typically so astounding that many people cannot even imagine it to be caused by processes or events that involve nature: yet the miracle took place in nature's physical world. Miracles, by their causal incomprehensibility, seem to demand the intervention of a superintelligence with infinite knowledge and infinite capabilities. Some miracles have caused nonbelievers to convert to a belief in God. Even experienced medical doctors have taken up faith that previously had worn away to the point of all but disappearing. Men and women have joined the holy orders because of a miracle's effect on them. Many miracles prove so astounding that they have shaken the faith of atheists and caused instant epiphanies by their implications about God, the saints, and the afterlife. One definition of a miracle is "an incredible effect without a physically known cause." Mirriam-Webster's definition of a miracle reads:

An extraordinary event manifesting divine intervention in human affairs.[460]

Stephen Greenblatt, winner of the 2012 Pulitzer Prize for General Nonfiction and 2011 National Book Award for Nonfiction, writes that miracles can have a secular appearance and meaning to remove it from religious interpretations. The secular miracle will still be a miracle, just not thought of as supernaturally caused; it will have no known cause and effect in nature, but will be attributed a cause to suit the religion of atheism: God is merely replaced with the unseen interactions of nature; there will be no independently observable or repeatable causal model. God is merely replaced by another god called nature. Such a secular definition for a miracle was given by Lucretius, about 100 BC. The secular miracle is given the name of a "swerve," meaning:

An unexpected unpredictable movement of matter.[461]

This definition is removed from nature, independent of it, and removed from science. The secular meaning of a miracle is given to remove it from any master plan or divine architect. The "stuff" of the universe that is the "cause" responsible for creation by the swerve is an infinite number of atoms moving randomly through space, like dust motes in a sunbeam, colliding, "hooking together," forming complex structures, "breaking apart" again in a ceaseless process of creation and destruction.[462] This definition of the secular miracle, a swerve, is as laden with as much intense belief and faith as found in almost any definition of a miracle. Both definitions link cause and effect with testimony. Both are parts of different religions. Neither is science.

The testing of miracles is exclusively a testimonial process and it is called "testimonial testing." However, the *effects* are the same for every creation model: natural selection, special creation, "use and disuse," and orthogenesis are examples that share the identical effects. There is one significant difference between secular and theistic creation models, however. Gradual secular creation models are missing the parts of creation that reveal that it ever took place: there are no incomplete creatures that are mandated by gradual creation (by any cause). Testimonial models do not show the physical causes in nature, which must be shown to be eligible to even be considered as possibly being science.

Upon reviewing the method of testing in this chapter, one finds miracles and natural selection to be the same type of testimonial models: both are inoperative in nature and both require testimony to stir beliefs, fuel the imagination, activate faith, and allow the model to become a means through which testimony about creation and evolution takes place. This is why evolutionary models are held tentatively—they are not rooted in nature, have no repeatability, do no foretelling, and are subject to rebuttal and changes of worldviews yielding different answers, making them subject to change. Darwin had long tried to have his readers change their "way of thinking" to his way, meaning that special creation was to be rejected. Unbeknown to many of those who do reject special creation, they are also rejecting the Bible and the God of Genesis, the God who communicates with mankind, performs miracles, and creates life, as shown in chapter 4, "Theistic Evolution."

Before a miracle is accepted,[463] it may be tested repeatedly over years by many different people, including medical boards of doctors. The many levels of each test, if finally deemed successful, will show cause linked to effect and the existence of a miracle. One may wonder, if testing is part of the scientific method, does successful testing mean that a miracle constitutes science? If not, how does the testimonial testing of natural selection differ from the testing of miracles? If testing natural selection shows it to be "science," then testing miracles should as well; this would mean that either both are science or neither is science because both are testimonial tests, with the only difference being the worldview being used: however, worldviews are not science. Further, if successful testing does not reveal miracles to be science, then what does the role of testing in the scientific method reveal for evolutionary biology models of creation? It should become apparent that testimonial testing does not reveal the same type of "cause and effect" as foretelling testing that reveals nature's operations and science. With testimonial models, only a faith-based world-view is advanced, not nature's operations or science.

The term "miracle" is derived from the old Latin word "*miraculum*," meaning "something wonderful," a striking interposition of divine intervention by a supernatural being in the universe.[464] The reader should recall that creation by "swerves" is synonymous with creation by miracles, with a difference only in the cause but both involving religious faith. With miracles, a physical event observed in nature, such as an unexplainable healing, is said to be caused by God. With swerves, the physical event is still unexplained, but assigned a different model (cause) of atoms that bounce around in some undefined and unseen way, with unknown processes, unknown rules of operation, and unknown relationships.

Some may say miracles occur when "the ordinary course and operation of nature is overruled, suspended, or modified."[465] However, to claim that the ordinary rules of nature are overruled is to claim that they are known and observed to be overruled. Without a causal model that is known to operate with nature's foretelling cause, it is impossible to know whether they have been suspended or overruled.

Testing of natural selection is performed using "testimonial tests," which are very unlike the foretelling tests involving nature and science. The testimonial tests readily identify the testimonial characteristics of the model involved. The tests show that the cause of miracles is God, for those accepting the tests. With miracles, as with natural selection, the tester testimonially attributes the "cause" by inference, logic, extrapolation, correlation, or some other testimonial means. In this same way as miracles, natural selection is attributed to being the cause of creating new creatures. Multiple sources for miracles are difficult to find, even in books dedicated to miracles. However, any specific miracle is not the important point as many miracles exist. The important point to consider is the manner by which a miracle is pronounced – by testimony, just like all evolutionary biology pronouncements for creation and natural selection. Natural selection's variations that accumulate into new creatures are also miraculous as variations are never observed accumulating into new body parts and new creature's bodies; the details of nature and the steps involved in creation are as unseen as they are with miracles. Testimony and arguments abound with Darwin and today's authors, with no nature or science in the operations of the creation models. There is no cause and effect of nature, there is only "testify and effect". That is a major point with miracles and their being shown with natural selection in this chapter: testimony abounds, not nature or science. Nature is observed only in the "effect," not the "cause." Three tests[466] of miracles are discussed next: the burn victim miracle, the stopped heart miracle, and the cancer patient miracle.

Miracle One: The Burn Victim Miracle

The subject of the first miracle is a Spanish woman whose entire body had been covered by third-degree burns of the worst possible nature. Photographs had shown her as basically a "lump of raw flesh." At the hospital where she was taken to die, attending physicians told the woman's relatives that they could only relieve the woman's pain until her hopeless struggle to live had ended. However, one of the relatives placed a picture of a holy man, along with a relic, on the dying woman's bandages. Overnight, the burns healed completely, without scars. One of the reviewers of the case acknowledged, "Simply by what we know about the multiplication of cells, we can say that this is beyond natural law. And in this case, the [new] skin was not only unscarred but like that of a newborn baby."[467]

Several doctors at the hospital renewed their religious practices after witnessing these events. This event seems to reveal that the known processes of burn healing and recovery did not operate and the known laws of healing were suspended, making the recovery a miracle by definition. That no burn scars remained further emphasized the fact that the process by which healing took place fell beyond known healing processes. In effect, the normal physical laws of healing were apparently overridden, demonstrating that a miracle had taken place. It may be claimed that laws not normally present in nature were somehow made active. Miracles are differentiated from science by their testimonial link between cause and effect, not by their simply being called "miracles" or the cause being God; the observed "astounding healing" is the effect. With testimonial testing's causal link being someone's testimony, one may accept the cause and effect—or reject it. This same type of link is used with natural selection, showing the model and the link to be no different in kind.

Miracle Two: The Stopped Heart Miracle

A doctor died; his heart stopped for two and a half hours. Then he came back to life after prayers for intercession, without any trace of brain damage or other residual effects. The attending physicians, some of whom were colleagues of this patient, agreed unanimously that the event could not be explained.[468] They could offer no account of healing using the known laws of physical biology or nature. As with the burn victim, the normal physical laws of nature remained unknown, with the result being that the event was a pronounced miracle. Testimony linked the cause of the healing and the healing itself (i.e., the effect). Faith also played a part in the causal link as it involves the four major parts of religion: origin, purpose, morality, and destiny. Any parts of nature involved remained unknown. For many, no part of nature was involved—only God.

Miracle Three: The Cancer Patient Miracle

A young mother in Australia had leukemia so advanced she had been given no more than three weeks to live. After her entire family prayed for the intercession of a miracle, the woman recovered completely within a few days. The chief physician on her case refused to believe that the cancer was in full remission and insisted she remain under observations. Fifteen years passed before doctors could agree that her healing had been both complete and beyond "scientific explanation."[469] The processes of nature that had been involved in her case remained unknown. No account could be offered of her healing using the known causal models of nature. As with all miracles, the link between cause and effect was testimonial and did not involve known parts of nature.

Summary: Testing Miracles

Miracles are but one causal model that depends on testimony: natural selection is another. Testimony can be expressed using extrapolation, correlation, guessing, fiat, abductive inference, personification, analogies, need, logic, analogies, arms races, adaptive landscapes, or other types of "tools" *that are causeless*: one may use them, but without showing any cause in nature, thereby disqualifying them from being science. No form of testimony shows nature's cause and effect; it can, at most, show "testify and effect," which may change a person's views, thinking, and religion.

One purpose of a "testimonial test" in evolution is to attribute or infer natural selection. That is the end result and purpose of using "historical inferences" to discuss evolutionary biology: one merely presents his or her worldview and beliefs in the form of inferences about fossils, claiming that natural selection or other evolutionary biology is responsible and that it is "science." There is no "cause and effect" involving nature, hence there is no science. In all cases, testimonial models and testimonial testing lead to a "way of thinking." The end result for testing miracles mimics the end result for all testimonial tests, including those for natural selection: to examine the link between cause and effect, such as inference or correlation, and reject or accept that link.

One characteristic of testimonial testing is that the *same test* can succeed for one person and fail for another, showing that nature is not involved, but the person's views are. If a person agrees with the testimonial test results, the test "succeeds." If a person disagrees with the test results, the same test "fails." Debating tactics, not science, are used to resolve differences. Regardless of the result, one fact is immediately apparent: the tester knows no more about any processes of nature involved with miracles (including creation) after using the model than before; that person must simply believe or disbelieve in the testimonial link. Nothing of nature's physical world exists in a miracle other than the effect; even the link between cause and effect, which occurs through testimony, is not part of nature's physical processes. This is especially true for natural selection.

An important aspect of miracles is that the testimonial link allows one to show his or her views about God and creation; miracles serve as the vehicles for that testimony. One may ask, if a miracle is successfully tested or if one holds it tentatively, is it now science? There is a biased standard on the part of natural selection supporters that allows testimony as a link between cause and effect for natural selection but disallows it for *other* faith-based systems, such as Genesis; this Darwinist view has all the hallmarks of a double standard. Natural selection has no more physical parts of nature to it than does a miracle.

Testing Natural Selection

This second part of the chapter shows that testing natural selection is like testing miracles; hence, it is being included here. Darwin's circle of friends included dissenters who did not accept his model of natural selection. Charles Lyell never agreed with natural selection, yet he was a significant mentor to Darwin. Thomas Huxley also never accepted natural selection, yet he defended Darwin. The person who departed significantly from Darwin was Alfred Russel Wallace, natural selection's co-discoverer. In the April 1869 issue of *Quarterly Review,* Wallace writes that "[A] Higher Intelligence has guided the same laws for nobler ends."[470] Darwin would naturally disagree, but natural selection was silent and inoperative in nature, leaving only Darwin's testimony against Wallace's: two opposing answers for the same model, both acceptable to the model of natural selection. Wallace writes:

> Such, we believe, is the direction in which we shall find the true reconciliation of Science with Theology on this most momentous problem. Let us fearlessly admit that the mind of man (itself the living proof of a supreme mind) is able to trace, and to a considerable extent has traced, the laws by means of which the organic no less than the inorganic world has been developed.[471]

To the consternation of Darwin, Wallace proclaimed the very thing that Darwin opposed: God having a direct role in creation. Today's supporters of natural selection also disagree because they support another religion. The tests shown here should be discussed by today's supporters as their analysis would be interesting and instructive—but not in any way they would like it to be, for religion is involved in all testimonial answers. Again, we have two opposing answers for the same model of creation that is said to be science by many of its supporters. Wallace continues with his view of God's involvement, which he held to be science:

> But let us not shut our eyes to the evidence that an Overruling Intelligence has watched over the action of those laws, so directing variations and so determining their accumulation, as finally to produce an organization sufficiently perfect to admit of, and even to aid in, the indefinite advancement of our mental and moral nature.[472]

In the above quote, we see Wallace's view of "direction" and "accumulation" being directed to form accumulations during the creation of man's "finer qualities." As Darwin had declared with the animal breeders, Wallace declared the use of intelligence, purpose, plans, and goals: the two men agreed with the use of those qualities in creation, with Darwin never showing how those qualities operated in nature. These two men displayed the glaring characteristics of testimonial models: the worldview of the user determined the very nature of natural selection's cause and effect relationship, and the words used in the model are not anchored to nature. In this case, the two views contradicted one another. Darwin held his view to be true while disagreeing with Wallace; Wallace held his view to be true while disagreeing with Darwin. In place of cause and effect in nature, their models show "testimony and effect." The link between cause and effect operates by testimony, using terms independent of and inoperative in nature. In reality, what is shown by a testimonial model is the worldview of each person using it, not the operations of nature—and not the characteristics of nature or science.

In Wallace's 1869 article, "The Limits of Natural Selection as Applied to Man,"[473] Wallace used tests to show that a creation model other than natural selection was responsible for the creation of a number of human features. Two types of tests illustrate the drastic and contradictory differences between the two types of testing: foretelling tests used for Newton's model of gravity and testimonial tests used for miracles, special creations, and evolutionary biology's natural selection. During Wallace's lifetime, he was widely acknowledged

217

as the co-discoverer of the theory when, in fact, he published about evolution before Darwin (Sarawak law). In fact, natural selection was often called the Darwin-Wallace theory, and the highest possible honors of the land were bestowed on Wallace for his role as its co-discoverer. These include:

- The Darwin-Wallace and Linnean Gold Medals of the Linnean Society of London
- The Copley, Darwin, and Royal Medals of the Royal Society (Britain's premier scientific body)
- The Order of Merit (awarded by the ruling monarch as the highest civilian honor of Great Britain)

Wallace was noted as being a clear and capable writer and an excellent field researcher. Only with the modern evolutionary synthesis of the 1930s, 1940s, and 1950s did natural selection became the widely publicized and debatably "accepted" mechanism of evolutionary change. The history of the events of discovery is largely forgotten (there was a new generation of biologists), and when interest in the theory revived, many wrongly assumed that the idea had first been published by Darwin in his book *Origin of Species*.[474] Wallace had been pushed to the side (as he was pushed aside by Darwin in the 1859 *Origin of Species*), and Darwin was made the major focus of natural selection largely due to the 1930s–1950s synthesis. Darwin is now largely credited for natural selection. In one author's words:

> Thanks to the "Darwin Industry" of recent decades Darwin's fame has been rising exponentially, overshadowing the important contributions of his contemporaries, like Wallace.[475]

Wallace's contributions to natural selection and evolution are rapidly surfacing. Wallace conducted a number of tests that showed limits to what natural selection could do. Natural selection's tests are all testimonial, including Darwin's. Natural selection does not contain the very things that would have allowed it to create the different bodies of each new creature, such as new chemistry, electricity, hormones, biology, shapes, organs, and architectures. Without the representation of the parts of nature that do the creating in natural selection, the model's capability for creation doesn't exist; it has no relationship or operating capabilities in nature. Without showing the parts of nature involved, it cannot even be considered science. With or without testimony, it cannot pass one test that involves nature's physical cause and effect.

The conclusions taking place in a person's mind form the proof, evidence, and explanations of natural selection, as is the case with all testimonial tests. Arguments form the proof of Wallace's conclusion that natural selection or God is responsible for creation. Such an approach essentially mimics that of a missionary, not an engineer, researcher, or someone working to achieve the type of knowledge possessing the characteristics of science. For Darwin, Wallace's conclusions did not represent facts, but for Wallace, they did. Both men saw facts through conflicting worldviews (Wallace's worldview including God and Darwin's worldview excluding God), rendering different answers to the same observations. For Wallace, God was not mystical but very real, existing in a different dimension or domain. The determination of what is or is not a fact has carried into modern discussions about natural selection. The National Academy of Sciences (NAS) defines "fact" thus:

> Fact: In science, an observation that has been repeatedly confirmed and for all practical purposes is accepted as "true." Truth in science, however, is never final, and what is accepted as a fact today may be modified or even discarded tomorrow.[476]

The NAS does not show how its definition is a scientific one. The carefully structured NAS definition of a *fact* is neither scientific nor correct. Facts are not determined by the number of people who believe them or hold them to be true. Facts are not considered "repeated" because different people see "testify and effect" or inferences in the same way. Facts are the same for every person, but beliefs that are used to determine "facts" may be different for different groups and persons. When using a worldview to make a testimonial determination, the "facts" are likely to change with a different worldview or when nature's facts contradict it.

A fact of nature is not held tentatively and does not need to be accepted as true. Examples include the temperature being 68°F, or water freezing at 32°F and boiling at 212°F (at atmospheric pressure)—for they are repeatable and the same for all people. As with all facts that are dependent on nature, they do not need to be "accepted as true," for they are dependent on nature and are the same for every person. Thus they do not waver by worldview and are, by definition, "true." One may merely repeat the observation of the fact, not conclude it from inferences that are causeless; conclusions inferred from historical or other observations are not "facts." Rabies germs identified by Pasteur and used in his vaccination models are "facts" because they are the same germs today as they were in his day. They will be the same germs forever and need not be held tentatively.

Successful foretelling models of nature are "facts" that do not change with time, such as those of Archimedes (buoyancy, lever), Newton (gravity, motion), Copernicus (sun-centered solar system, and Ohm (current, power). These facts are held not tentatively but absolutely in nature. Each of these models (facts) operates without testimony. However, "testimonial facts" do differ between people, do change with time, and must be held tentatively as they are worldview dependent, not nature dependent. Wallace's tests in this chapter clearly show why they must be held tentatively and why they depend on consensus and popularity to be called "facts": they are faith-based, as are all testimonial models that deal with man's origin, purpose, morality, and destiny; they are religious models. Three of Wallace's five natural selection tests include the following: test 1: brain size; test 2: man's naked skin; test 3: creation of the feet and hands of man.

Test 1: Brain Size Observation of the Brain of the Savage Shown to Be Larger Than He Needs It to Be [Sic]

Certain beliefs surround natural selection. One of these beliefs (considered to be "fact" by some) is that nothing can be created before it is needed. In this test, Wallace notes that, because natural selection does not create for future requirements, we must conclude that wherever we encounter creations with capabilities far greater than needed or used, natural selection could not have been the creation model. Wallace uses this tenet of natural selection to make observations about the size of the brain for "savages," which is larger than needed. Wallace bases his tests on accepted tenets of natural selection, shown in the following:

> [...T]he large brain he [the savage] actually possesses could never have been solely developed by any of those laws of evolution [by natural selection], whose essence is, that they lead to a degree of organization exactly proportionate to the wants of each species, never beyond those wants—that no preparation can be made for the future development of the race—that one part of the body can never increase in size or complexity, except in strict coordination to the pressing wants of the whole. The brain of prehistoric and of savage man seems to me to prove the existence of some power, distinct from that which has guided the development of the lower animals through their ever-varying forms of being.[477]

If we accept Wallace's views as correct, his tests show that these tenets surrounding natural selection are violated, leading to his testimonial conclusion that God created those human parts to meet future needs. Wallace states:

> [W]e must feel satisfied that volume of brain is one, and perhaps the most important, measure of intellect; and this being the case, we cannot fail to be struck with the apparent anomaly, that many of the lowest savages should have as much brains as average Europeans.[478] The idea is suggested of a surplusage of power; of an instrument beyond the needs of its possessor.[479] Hence, the brain was made larger than necessary to meet the needs of the "savage" or "primitive" man.[480]

221

Because he encountered "larger-than-needed brains," Wallace concluded, using inference, that natural selection could not have created them; something else was responsible for creation. The test is complete, with the following results:

- Wallace shows that the limit of natural selection was reached and demonstrates creation has taken place ahead of a current need.
- Natural selection could not have created the human brain, as the brain is larger than needed by "savages" and "primitive man."
- Another creation model that does create ahead of present needs created the brain. For Wallace, God was responsible. Hence, for him, this testimonial test proves that God, not natural selection, was the creation model.

Test 2: Man's Naked Skin

In this test, Wallace writes that human's naked skin could not have been produced by natural selection. He writes:

> It seems to me, then, to be absolutely certain, that natural selection could not have produced man's hairless body by the accumulation of variations from a hairy ancestor. The evidence all goes to show that such variations could not have been useful, but must, on the contrary, have been to some extent hurtful. If even, owing to an unknown correlation with other hurtful qualities, it had been abolished in the ancestral tropical man, we cannot conceive that, as man spread into colder climates, it should not have returned under the powerful influence of reversion to such a long persistent ancestral type. But the very foundation of such a supposition as this is untenable; for we cannot suppose that a character which, like hairiness, exists throughout the whole of the mammalia, can have become, in one form only, so constantly correlated with an injurious character, as to lead to its permanent

suppression—a suppression so complete and effectual that it never, or scarcely ever, reappears in mongrels of the most widely different races of man. Two characters could hardly be wider apart, than the size and development of man's brain, and the distribution of hair upon the surface of his body; yet they both lead us to the same conclusion—that some other power than natural selection has been engaged in his production.[481]

Wallace successfully completed his test. The lack of hair on the human body shows it was "to some extent hurtful"[482] and thus "suppressed" in humans. As Wallace deemed this characteristic injurious, some other creation mechanism must be responsible; for Wallace, God was responsible.

Test 3: Creation of Feet and Hands of Man

Wallace provides this test to show that natural selection could not have created human hands and feet, thus showing that some other creation model must have created them (an "intelligent power or Intelligent Power"). In Wallace's words:

The specialization and perfection of the hands and feet of man seems difficult to account for. Throughout the whole of the quadrumana the foot is prehensile; and a very rigid selection must therefore have been needed to bring about that arrangement of the bones and muscles, which has converted the thumb into a great toe, so completely, that the power of opposability is totally lost in every race, whatever some travelers may vaguely assert to the contrary.[483]

In this test, Wallace finds it difficult to understand how selection could have taken away something very useful, such as prehensile power (meaning that they could be used for grasping by wrapping around an object). He does not think that selection would commit such a creative act. Wallace writes:

It is difficult to see why the prehensile power should have been taken away. It must certainly have been useful in climbing, and the case of the baboons shows that it is quite compatible with terrestrial locomotion. It may not be compatible with perfectly easy erect locomotion; but, then, how can we conceive that early man, *as an animal,* gained anything by purely erect locomotion?[484]

The hands had capabilities unused by savages and early humans or ruder predecessors; they were created before they were needed or used and thus could not have been created by natural selection, as it does not create ahead of need or use. Another tenet is that natural selection does not have plans or goals. Wallace writes:

Again, the hand of man contains latent capacities and powers which are unused by savages, and must have been even less used by palæolithic man and his still ruder predecessors. It has all the appearance of an organ prepared for the use of civilized man, and one which was required to render civilization possible.[485]

If natural selection is not responsible for creation, then it must be some other creation model. Wallace did not think the name of that power of creation was important, whether God, intelligent force, or some other name. He writes of the intelligent power:

Apes make little use of their separate fingers and opposable thumbs. They grasp objects rudely and clumsily, and look as if a much less specialized extremity would have served their purpose as well. I do not lay much stress on this, but, if it be proved that some intelligent power has guided or determined the development of man, then we may see indications of that

224

power, in facts which, by themselves, would not serve to prove its existence.[486]

This test was successful in Wallace's mind no differently than Darwin thought his "tests" were successful. As the hand's capabilities went largely unused by early humans as well as "savages," it had capabilities created ahead of being needed or used. In this way, Wallace showed that natural selection could not have been the creation model: test complete. For Wallace, God was responsible. For Darwin, God was not responsible. Here we have another complete contradiction by two men using the same testimonial model: natural selection, showing its worldview dependency and complete removal from science.

Summary: Testing Miracles and Natural Selection

In addition to the above three tests of natural selection, Wallace also conducted two other tests. His fourth test was called "The Voice of Man Was Not Created by Natural Selection,"[487] and the fifth was "The Origin of Some of Man's Mental Faculties by Natural Selection Is Not Possible."[488] These tests are of the testimonial type and are perfectly legitimate, but not as science; they are no different in kind than Darwin's tests and are as much "science" as Darwin's in that they "pass tests."

For 150 years, the public debate has been about science versus religion. Natural selection is incorrectly said to be science. As portrayed in Genesis, miracles and special creation are correctly said to be religion. But the public debate always consists of arguments, not science. All of evolutionary biology consists of arguments formed from inferences and creation models that are testimonial, possessing none of the characteristics of science. Whether finch beaks, peppered moths, or mimicry, the tester forms the conclusions about causal links and argues their case. The tester determines what is successful or a failure as illustrated with the above tests of miracles and natural selection. Nature, in the form of natural selection, is inoperative—always on the sidelines, literally doing nothing, and empty of nature's infinite parts, rules, and relationships. Wallace's (or Darwin's) tests consistently show one thing: there is no science involved with testimonial tests or with the testimonial model.

Such tests and attributions are the identifying hallmarks of religion and are the antithesis of science—its exact opposite. With natural selection, as with miracles and special creation, you are free to accept, reject, and argue the test results or uses of the causal model. A complete worldview dependency exists with all testimonial testing that forms the basis for choosing each success or failure, which is not decided by the testimonial model itself, for it is incapable of relating to nature.

Samuel Butler, in his 1890 publication, *Evolution Old and New*, writes:

> [T]rue theories make themselves, they are not made, but are born and grow; they cannot be stopped from insisting upon their vitality by anything short of intellectual violence,[489] nor will a little violence only suffice to kill them. True theories [models] are but the spontaneous mental coming together of facts, which have combined with one another by virtue only of their own natural affinity.[490]

It is not difficult to see Butler's emphasis on "mental" models, sometimes called paradigms, and their use; that is, he emphasizes testimonial models, not those with nature's cause and effect. The "insisting upon the vitality of a model" takes place in the mind with worldview filtering, because nothing short of intellectual violence will take place if one rejects the model. Testimonial models are all about "how one thinks" about nature and creation. The users of the model determine the model's attribution to account for, explain, or prove the observations are caused by their model. Testimonial models are belief-system dependent.

In contrast, science has no such worldview dependency. What a person thinks, believes, and feels is not operative in science's foretelling models. This is why foretelling models do not conflict with religion, but testimonial models do (such as natural selection). Samuel Butler's spontaneous, mental "coming together of facts" is not nature-dependent but worldview- dependent. In 1882

Butler, who did not believe in the Bible or God, tells us why a test and its results can differ for two people. Butler writes:

> When a number of isolated facts...take form, group them-
> selves together coherently, and present the mind so vividly
> with an idea of their interdependence and mutual bearing
> upon each other, that no matter how violently we tear them
> asunder they insist on coming together again; then, and not
> till then, have we a theory.[491]

Butler is practically defining the term "worldview" and the ideas that agree with that view. Using that view as a filter, ideas and data may be selectively chosen while omitting anything contradictory. The only way to have facts "group themselves together" is to have an accommodating worldview. That they come together is a process of the imagination, not nature. In his 1802 book, *Natural Theology. Or, Evidences of the Existence and Attributes of the Deity, Collected from the Appearances of Nature*, Paley writes about worldviews (without using the term), how a person acquires a worldview, and how it reveals his or her character:

> [P]erhaps almost every man living has a particular train of
> thought, into which his mind glides and falls, when at leisure,
> form the impressions and ideas that occasionally excite it;
> perhaps, also the train of thought here spoken of, more than
> any other thing, determines the character of the man.[492]

A man's character and views allows for particular trains of thought to "glide and fall," naturally accommodating a worldview. In this single statement, Paley showed that a testimonial model reveals what is inside of the tester, how he or she views the world, or what he or she thinks of creation.

Testing alone, like testability or tentativeness, does not reveal what is or is not science. Testing reveals the characteristics of the model being tested, such as foretelling, confirming observations, independence of worldviews, and independent repeatability: these are some of the characteristics of science. Testing shows the model to either possess the characteristics of science or the characteristics of models belonging to religion: natural selection belongs to a religion called Darwinism. The characteristics of the model *lacking* nature's physical components show all that is necessary to determine that natural selection will never be the type of knowledge that is called "science." Natural selection stands in complete contradiction to Newton's, Pasteur's, and Archimedes' foretelling models that *are* science, for it depends entirely on the worldview and faith of the tester to link creation to evolution. Natural selection's inoperability in nature does not matter to those who support it because, for them, it serves the greater purpose of being a platform for speaking against special creation and all other competing models of creation and evolution. That is natural selection's only role—one that is not science.

CHAPTER 12

ZERO CHANCE

Introduction to Chance Creation of Variations: The First Part of Natural Selection

Chance creation of variations is the first part of natural selection's two parts. Darwin did not know how *chance* operated. In the 1859 *Origin of Species*, he writes that chance is a name for ignorance:

> I have hitherto sometimes spoken as if the variations—so common and multiform in organic beings under domestication, and in a lesser degree in those in a state of nature—had been due to chance. This, of course, is a wholly incorrect expression, but it serves to acknowledge plainly our ignorance of the cause of each particular variation.[493]

Ignorance cannot be part of a science model. Ignorance might be excusable if the accumulation of variations were commonly observed and all that was left to do was to find the cause. That was not the case. Continuing in the 1859 *Origin of Species*, Darwin embellishes on the ignorance surrounding chance:

> But we are far too ignorant to speculate on the relative importance of the several known and unknown laws of variation; and I have here alluded to them only to show that, if we are unable to account for the characteristic differences of our

domestic breeds, which nevertheless we generally admit to have arisen through ordinary generation, we ought not to lay too much stress on our ignorance of the precise cause of the slight analogous differences between species.[494]

Neither Darwin nor Wallace ever knew how many variations were in an accumulation or how many accumulations were needed to form any one body part or system of organs. They did not know how variations or accumulations related to each other, how accumulations physically related to new body parts, or how they accumulated to become adaptations. They did not know how variations worked within nature's rules or what the rules were during creation. Even their model of natural selection does not reveal anything about the creation process or what is being created. Those questions remain unanswered to this day because it is difficult to determine answers for something that is not defined or described in nature or in the creation model that is said to have created creatures.

Neither Darwin nor Wallace could ever show that any variations accumulated to form new creatures: they could only argue for their beliefs about what took place and hope the reader would think the same way. That accumulation process still cannot be shown today, even among the fossils where billions of years have passed and billions of creations have taken place, with many times that number that had to be created as "intermediates" that are incomplete during their entire existence between fully completed creatures—their parents and the ones they were supposed to become. Supporters of natural selection cannot show any creatures being formed between the starting parents and the final new creatures that were created, with one of each sex being created for reproduction. Somehow the two sexes' creations were coordinated for completion at the same time and geographic location, for every type of creature, to make reproduction possible for each kind. That "time-location-biology" coordinated creation of the two sexes is a "mystery" to Darwinists. It should be, for it is not science. Nowhere in natural selection is this mystery moved toward solution. Not even a hint is offered anywhere for how the creation of sex took place for each type of creature.

In theory, practice, and reality, chance creation of variations amounts to a mere words in natural selection that are empty of nature. When variations are said to be mutations, the term "mutations" becomes another undefined word that is also empty of nature: mutations, like variations, have no weight, no volume, no dimensions, no physical arrangements, no functions, and nothing that can be used to show they have a reality; this is not a type of term that is found in science. Mutations are said to be the "raw material" used to create intermediates and creatures in nature, at least by supporters of natural selection. This chapter will show that there is a *zero chance* of any two variations accumulating in a "direction" that will ultimately result in the creation of an adaptation or the successful creation of a new part of a new creature's body.

What Is *Chance* in Natural Selection?

The first part of natural selection's two parts is the chance creation of a new variation, more accurately called a new mutation. Even in theory, each chance mutation is likely to be unique. The reason for two variations not likely being the same is that every characteristic of a variation is determined by chance. One characteristic of a variation is its materials. Other characteristics are the materials' shapes, dimensions, and locations (of the variations in an accumulation).

The timing of a variation's creation by chance is never known and cannot be determined, resulting in the inability to know when, if, or where a chance creation of a variation is to take place. Most importantly, there is no way to determine that the variation that is "needed" is the one that will be created—or if anything is to be created at all. Natural selection does not shed any light on the creation of variations that are supposed to produce evolution. To create anything, a specific variation that is needed to have the accumulation form a body part must match perfectly with the accumulation to which it is being joined, characteristic for characteristic. If the variation is not perfectly aligned with the accumulation to which it is being joined (characteristic by characteristic), then the newly forming body part that the accumulation is forming fails. For example, the location of the nerves in a variation must align with location of nerves in the accumulation; they must match each other's materials, dimensions, and shapes. In another case, veins in a variation must align with veins in the accumulation and must also have matching materials, dimensions, and shapes; bones must align with bones; and body parts that are being formed must align with the body parts with which they operate.

Due to there being no observable physical population for determining what "chance creation" of a variation means in nature, creation is no different than rolling dice with no marks on them, which is playing dice with no known population. The fact that variations are not defined means they have no known physical characteristics. Such chance has more of the characteristics of a spirit than an operational part of nature.

There are other problems with determining the chance of a mutation's creation taking place. There is no means of determining the chance of genetic material being altered by mutation to create what is needed in the newly forming body. The creation process is not defined, has no rules, has no schedule, leaves no residual trail, and has no model of how it operates in a cause and effect manner in nature. One may look at any fossils and be unable to show variations, accumulations, or chance operating. Only inferences may be shown, which consist of personal testimony. This problem is intensified by natural selection offering nothing more than words that are empty of nature, offering no help in knowing what is taking place during creation. With natural selection, it is impossible to determine if anything is ever taking place in nature. One is forced to testify one's views in place of an operational causal model.

The Values of Chance

A probability (chance) is a number that ranges between zero and one. At a value of *zero*, an event such as creation of a variation is shown to be impossible. At a probability of *one*, an event will unquestionably take place; it is a certainty. An event with a probability of one half, or 0.5, means that a particular event will take place half of the time. For example, coin flipping has observable populations, rules, relationships, and procedures. The entire coin-flipping population consists only of heads and tails, each having a 0.5 chance of taking place. With all chance models that have populations, a pattern exists for all possible outcomes. Of one hundred flips of a coin, the pattern would show fifty would be heads and fifty tails, at least in theory.

The Types of Chance

At least four types of chance exist. The four types include gaming chance, actuarial chance, normally distributed chance, and emotive chance. The first three exist in nature's physical world because they have observable populations. The fourth one does not, because variations that accumulate to become body parts have never been observed and have no populations. Darwin's model of natural selection uses only the fourth one—emotive chance. It offers no means of determining how it operates in nature, as it has no population by which chance could be determined. That is, the intermediate creatures are never observed that are in the process of having their new organs gradually created, variation joined to variation. Emotive chance rests on how one thinks about what could have happened using inferences and a worldview that ordains those inferences. To compute probabilities using natural selection is a physical impossibility, for there is nothing to base the probability on. One has to imagine the physical events and populations of mutations taking place because they do not exist outside of the mind. This type of imagined chance has nothing to do with nature or science, but it is the one used for natural selection. Different values of chance may be obtained if one were to use an existing creature as a starting point and compute the chance of creation by imagining how creation took place gradually. When starting with single-celled creatures and computing the chance of mutations accumulating new body parts, perhaps to form an elephant, different values are obtained, even by different people. These are the intermediate creatures with bodies that are gradually being created by variation being added to variation, each variation being undefined and unobserved.

Variations' Characteristics

There are many characteristics that variations may have possessed, but only a few are considered here to illustrate how these characteristics relate to creation and evolution by natural selection. Materials are the first characteristic of a variation. Many different materials are needed to create cells, tissues, skin, muscles, nerves, organs, and every part of the body. The second characteristic of a variation is the material's shape. Shapes are necessary to form the different parts of each body, such as cell walls, skin, ears, fingers, hearts, legs, arms, bones, and so on. Dimensions are the third characteristic of a variation and must be specified by the creation model. Bones, for example, have varying dimensions as they are not uniform in their shapes. The fourth characteristic of a variation is its location within an accumulation. Each variation has a location in an accumulation, and each accumulation has a location in the body. These four characteristics are created simultaneously. Other characteristics that that are not considered here may be: alignments; joining method to form accumulations, body parts, or adaptations; strength of each variation and body part; sequences of creation; and the architectures of body parts and creatures' bodies.

Components within a Variation

Variations are not necessarily homogeneous. They do not necessarily consist entirely of one material. Variations may have embedded components that consist of parts of nerves, tissues, capillaries, veins, arteries, organs, muscles, glands, fluids, and so on. Examining any cross section of a body reveals that it contains skin, nerves, glands, veins, arteries, and organs: these are the components in the cross section and they may be the components in a variation. As variations accumulate to form completed body parts, the components must also accumulate or creation fails. How such accumulations are created, or what they physically are, is not shown in natural selection, leaving the accumulation process undefined, unobserved, and a mystery. Testimonies about variations merely assume that "variations accumulate" with details omitted; miracles operate in this manner as well—details omitted.

The Chance of Creating a Mutation's Materials, Shapes, Dimensions, and Location Is Zero

Among the first characteristics of a newly created variation (mutation) are its material, shape, dimensions, and location. These four characteristics are created simultaneously, as are each of its embedded components. Consequently, if a nerve is a component in a variation, then the accumulation to which it is joined must have a nerve as a component as well, with the same material, shape, dimensions, and location; otherwise, creation will fail. These four characteristics did not exist anywhere on earth or anywhere in any newly forming body until they were mutated into existence from the first living creatures—something created from nothing, so to speak. Each variation and its components may be made of randomly generated materials for bones, glands, nerves, electrical signals, neurochemicals, acids, blood, and hormones and must match the variations to which it is joined. That process is unobserved but is inferred to have taken place. Inference is not science and is not causal.

Problems arise by the fact that each variation and its material could be randomly created anywhere in the body without regard to what is being created. For example, bone material could be created to form part of the eye lens, thus destroying the creation of the eye. Similarly, a variation with lens material could be created in the heart, thus destroying the creation of the heart. Any part of the body could be formed with any material: bones could be formed of muscle, thus destroying the ability to support the body. A random creation of a variation would have the "wrong" material being created for any part of the body, causing failure of the newly forming body. An endless series of variations' creations could cause countless numbers of mutations to take place and still not have a suitable material, shape, dimensions, or location created for the creation of any one specific body part. The skeleton could be created from many materials and thus not function to support the body because the skeleton need not be made of bone material. Because of the random location for each variation and the random location of each of its components within the variation, the chance of creating one variation with the necessary characteristics for accumulation to form one body part is one out of an infinite number of possibilities, which is a zero chance.

There is an infinite number of variation shapes that are possible, with any one shape having a zero chance because one divided by infinity is zero. Variations' dimensions are randomly created and may be any one of the infinite number possible, with the needed dimension being one chance out of an infinite number, which is a zero chance of a variation's creation for the one dimension that is needed for successful accumulation. The random location of a variation may be anywhere in the body, which contains an infinite number of possibilities, the chance of any one location being one out of an infinite number, which is zero chance for a successful accumulation of the variation and its components in any one location.

The chance of creation of a biological material may also be considered from a view of the natural selection's inherent flaws. No biological material is found as a part of the model of natural selection. This omission makes it impossible for it to have any role in the creation of materials no matter what their shape, dimensions, or where they were to be located. As a result of not having materials, shapes, dimensions, or locations or any other biological characteristics in natural selection, it could never be the model that represented nature's creation process for evolution, and it could never be a model that is science.

To illustrate the importance of the point that a model containing the parts of nature that are used to determine an effect using Newton's model of gravity, distance is one of the parts of Newton's gravity model. If distance were omitted from his model of gravity, the entire model would fail to represent what occurs in nature: Newton's model of gravity would fail. If a term that is independent of nature were substituted for "distance," such as "near" or "far," the model would again fail. The same failure would happen if Pasteur had omitted the rabies germ from his rabies vaccination model. Similarly, if Jenner had omitted smallpox or cowpox from his vaccination model, his model would fail. These models would not function in nature without each part being shown. Because of the absence of nature's parts from the model of natural selection, including bodily materials, it is inoperative in nature. It has a zero chance of representing creation of materials. It has a zero chance of creating anything.

Summary

Materials, shapes, dimensions, and locations are parts of nature and the physical world; they are characteristics that must be shown in any model of creation if it is hoped to operate as a science model that causes creation. These parts of nature (and all other parts of nature) are missing from natural selection and every other model of creation that ever existed in evolutionary biology, such as "use and disuse," orthogenesis, or special creation. Darwin described variations as beneficial (good) or harmful (bad), but he never defined what variations were or what those terms meant in nature. Darwin never observed good and bad variations. He never observed "bad" variations being "rigidly destroyed."[495] He used this vagueness of variations' definition as a springboard for testifying that chance creation took place. That vagueness of the definition of variations and mutations still is used today. Vagueness is not science, and nature is not vague.

Variations accumulating to form body parts have a zero chance of taking place for several reasons. One reason is the absence of nature in natural selection: there is nothing to accumulate except testimony. The model that is supposed to be responsible for creation and evolution cannot show anything taking place; hence, natural selection has a zero chance of creation or evolution. Next is the random creation of variations and their components that cannot be moved once created. Being immovable upon creation mandates random variations cannot be "selected" or "directed", making creation an entirely random process with no possible means of selection or direction having any role except testimonially. The immediate physical fixation of variations upon creation, within the body or accumulation, prevents any possible alignments of variations forming body parts, which dictates another zero chance of gradual creation or evolution; variations cannot be moved once created. Finally, chance in natural selection is emotive, existing without a population, making it removed from nature and science. Each of these methods shows natural selection has a zero chance of variations accumulating to form body parts. In confirmation of there being a zero chance of creation by natural selection, there are no gradually forming incomplete intermediate creatures in the fossils. What is shown in the fossils are fully completed creatures, with all intermediates being fully completed as if by saltations. The only place where evolution by natural selection has no problems with creation and evolution is in the biology and science textbooks, where creation is always a dogmatic certainty. That this is a religious certainty is not openly stated.

CHAPTER 13

WORLDVIEWS

Introduction

Imagine looking through one telescope and seeing one thing and then looking through another telescope *at the same thing* and seeing something different. Worldviews can be like that. Worldviews are more than how you look at the world; often, they are a projection of yourself and your character onto the world, filtering observations to provide the very vision you use to form ideas about the world. Facts become what you believe them to be (*credendo vides*). You imprint them onto the world, and they become the "facts" you "observe." Over the centuries, worldviews have always existed, invisible prior to their discovery. New worldviews that were made visible by their discovery include germ theory, circulatory systems, synthetic biological materials, vaccines, genetics, the sun-centered solar system, atoms, gravity, field theory, and relativity. Each was added to our view and understanding of nature's operations through the discovery of foretelling models that used these new worldviews. Each new worldview existed, waiting to be observed, but had not been discovered to that time.

The same holds true today: there are new worldviews waiting to be discovered. Holding one worldview may likely block the discovery of a new one that would allow us to create new, undiscovered models of nature. Newton's model of gravity changed the mechanical worldview. Faraday's electromagnetic fields worldview was added to the existence of nature. Pasteur's vaccination model solidified germ theory and solidly introduced modern vaccinations. Einstein's worldview made relativity visible when others could not see it. Hence, the past discovery of new worldviews has shown us that it is a mistake to think that the powers of mere observation show us all there is to see. With a new worldview, what one person cannot see today, another can see tomorrow—very clearly.

The world around us always contains discoveries that lie unseen until the "right person" comes along. In each case, someone opens the door to the existence of a new worldview that changes how we view, think, and talk about nature. This is because worldviews are parts of nature ready to be observed and used but remain unseen until their discovery. These new worldviews often accompany unprecedented ways of developing models, forming revolutions in how models are developed and portrayed. Worldview revolutions enable the creation of new types of models that initiate ways to use nature, sometimes called scientific revolutions; perhaps worldview revolutions are more to the point.

Worldview revolutions also enable revolutions in faith-based models. A person's worldview determines what he or she sees by acting as "worldview-colored glasses"; it determines how a person thinks and, therefore, what he or she believes. Seeing a new worldview may bring about an epiphany. It is the means by which you understand the world around you and the way you feel about it. New worldviews made many new advancements in understanding nature possible, such as Pasteur's use of germs for vaccinations and Newton's "action at a distance" for gravity. The origin of the term "worldview" is a literal, word-for-word, root-for-root translation from the German[496] word "*Weltanschauung*," meaning "a look onto the world." It implies a concept fundamental to German philosophy and learning and refers to a wide world perception. The way one "looks onto the world" provides a means of seeing, learning about the world, and believing. Every person simultaneously uses many worldviews to understand his or her environment and how nature operates; he or she may even use a faith-based worldview, such as atheism, Christianity, Judaism, or another religion to understand the world.

A worldview may include God or the denial of God's existence. The battles between those two faith-based worldviews continues to this day, especially in the realm of evolution and evolutionary biology. As we have no way to prove the existence or nonexistence of God using nature's involvement, that difference will continue into the unseen future, accompanied by the reasons people hold their views. The worldviews come from within, not from without, just as with natural selection. A worldview may contain borders outside of which nothing is thought to exist. Secularism is such a worldview, which holds that nothing exists outside of what we detect with our senses. It is a filtering worldview that denies any thinking contradicting the tenets that are part of it. One cannot hold a theistic worldview simultaneously with an atheistic one and be consistent, as the chapter on theistic evolution illustrates. A Descartes whirlpool gravity model cannot be held simultaneously with a Newtonian model of gravity. A sun-centered solar system cannot be held simultaneously with an earth-centered one. Though not always, worldviews are often exclusive.

One worldview may prevent seeing other conflicting worldviews, or even the possibility that they exist. Thus, discovering new ideas is not likely to take place if that discovery involves the use of conflicting worldviews. For example, the mechanical worldview would not allow the creation of models that showed Faraday's field theory; the worldview that denied the existence of germs would not allow the creation of models for vaccinations. No concepts existed in the mechanical worldview for forming models of "fields" that Michael Faraday discovered in 1831.[497] Nature contains both worldviews—mechanical and fields—but some holders of one denied the other.[498] To portray observed processes more completely in nature, you must use both views at the same time: mechanical and fields. When you enter the "field worldview," you enter another way of looking at the world of nature that supplements existing worldviews. Successful models of nature come from the "fields" worldview; it successfully portrays nature's processes using successful models that relate nature's cause to effect with fields that are impossible to conceive or use in the mechanical worldview. The following account, taken from Lovejoy's classic book, *The Great Chain of Being*, describes how a worldview operates:

> [A worldview operates with]...implicit or incompletely explicit *assumptions*, or more or less *unconscious mental habits*, operating in the thought of an individual or a generation.

It [the worldview] is the beliefs which are so much a matter of course that they are rather tacitly presupposed than formally expressed and argued for, the ways of thinking which seem so natural and inevitable that they are not scrutinized with the eye of logical self-consciousness, that often are most decisive of the character of a philosopher's doctrine, and still oftener of the dominant intellectual tendencies of an age.[499]

Every researcher, investigator, and worker uses one or more worldviews to create the conceptual models that portray how they think nature physically operates. Some of them prove useful in accurately portraying nature. Every religious person uses a worldview to make observations of the world, including those of the atheist religion. That is, a religious worldview explains, accounts for, gives evidence of, and even proves a religion to be true, at least for the person accepting it. Two conflicting religious worldviews observe the same world and see each of their views proven to be true. Some religious worldviews include actively eliminating other competing views, which is true of the Darwinism worldview, seen from reading "evolution by natural selection" evolutionary claims. Every person of every background uses different worldviews in his or her daily life, whether for modeling nature, using nature, or using beliefs. Some of the major worldviews include the mechanical worldview; the "fields" worldview; the spontaneous generation worldview; and the Darwinist worldview. A short description of each follows.

1. The Mechanical Worldview

The mechanical worldview includes only force, distance, matter, and time. These menu items serve as a pallet from which to form models of how nature operates, and they are all that existed in the mechanical worldview, making it a small pallet. The mechanical worldview did not allow changes to some characteristics; for example, a distance of one foot was always one foot, never to be changed; one second of time was always one second of time and did not shrink or grow; one pound of force is always one pound of force and did not change. This worldview received a shock with Einstein's relativity worldview, where no such absolutes exist, and distance and time could change by expanding or contracting.

Galileo and Descartes believed that the world was made of matter in motion. They thought that all objects were composed of particles or atoms and that these interacted in accordance with fixed, natural laws. This view of the natural world became known as "mechanical philosophy."[500] A drawback of the mechanical worldview or any worldview is that it may not suffice to answer questions or pose solutions for the problems found in nature. The mechanical worldview was not sufficient to compose models needed to answer important questions. When groups form around one particular worldview, they sometimes block changes, as is the case for many worldview changes. For example, a "flat earth" worldview blocks one of the round earth; a "germs do not exist" worldview blocks the use of the germ theory worldview.

Mechanical philosophers based their models and view of the world on machines, not organisms. They wished to produce general models that accounted for, in quantifiable terms, many different types of interactions observed in the world. They even held "thought" to be simply matter in motion.[501] The idea that human bodies were "possessed" by "incorporeal spirits" such as souls was unacceptable to some people who used a mechanical worldview; those people refused to accept the existence of other possible worldviews. The mechanical worldview was never erased by any other new view of the world. Field theory and its accompanying worldview were later discovered by Faraday, but existed side by side, with the mechanical worldview adding capabilities that could not be obtained solely from the mechanical worldview. Einstein's relativity worldview never erased Newton's worldview. Newton's worldview is used to this day as his models are highly accurate at speeds well below that of light. A synergy was obtained using both worldviews, where the sum proved greater than the individual parts.

2. The Fields Worldview

The concept of fields (as in electromagnetic fields) was totally foreign to the mechanical worldview. Michael Faraday discovered his model of induction, called the "law or model of induction," in 1831. Many in his time tried to use the mechanical worldview to develop models that could portray fields and thereby show the way nature's cause and effect operated. No models proved successful that portrayed fields using the mechanical world of matter and forces. When Faraday first conceptualized the "field" view of the world, it was used it as no more than a means of facilitating the understanding of phenomena from the mechanical point of view.[502] The recognition of the new concepts grew steadily, until a new way of looking at nature and its processes was born that could not be imagined using the mechanical view; a new reality was created. This illustrated the fact that one worldview (mechanical) often cannot show the processes in another worldview (fields).

The language of mathematics often makes a model more useful and accurate; mathematics allows easier foretelling, which is then followed by confirming observations. Despite some views to the contrary, Darwin's worldview or ideas were not needed or used to develop models or do research by Archimedes, Newton, Faraday, Maxwell, Kepler, Einstein, Pasteur, or anyone else. Darwin or natural selection is not needed for any aspects of science, technology, biology, or operations of nature. His and natural selection's sole use is to convince others about "how to think" or observe nature.

3. Spontaneous Generation Worldview: Life from Nonlife

It was commonly thought, as late as the seventeenth century, that sets of instructions (recipes or models) existed for creating life from nonlife by spontaneous generation. Some examples of these recipes follow:

> Take sweaty rags, wrap them around wheat, and set them in
> an open jar. In twenty-one days, you'll "create" mice.

> For rats, just throw garbage in the street. In a few days, rats
> will take the place of the garbage.[503]

All over the world, in Europe, Asia, Africa, and the Americas, mankind was formulating recipes for "creating" bees, lice, scorpions, maggots, worms, frogs, and so on.[504] In 1668, 191 years before Darwin published the 1859 edition of *Origin of Species*, Francesco Redi, an Italian scientist, designed a "scientific" experiment (a test) to disprove spontaneous creation of maggots by placing fresh meat in each of two different jars. He left one jar open; the other he covered with a cloth. Days later, the open jar contained maggots, whereas the covered jar contained none. He noted that he found maggots on the exterior surface of the cloth that covered the jar. Redi successfully demonstrated that the maggots came from fly eggs and thereby helped to disprove spontaneous generation.[505]

In England in 1775, John Needham challenged Redi's findings by conducting an experiment during which he placed a broth, or "gravy," into a bottle, heated the bottle to kill anything inside, and then sealed it. Days later, he reported the presence of life in the broth and announced that life had been created from nonlife. The test appeared to prove that life does come from nonlife, but in actuality, he did not heat it long enough to kill all the microbes, as was later proven by Italian scientist Lazzaro Spallanzani.[506]

Spallanzani (1729–1799) reviewed both Redi's and Needham's data and experimental design and concluded that perhaps Needham's heating of the bottle did not kill everything inside. He constructed his own experiment by placing broth in each of two separate bottles, boiling the broth in both bottles, then sealing one bottle and leaving the other open. Days later, the unsealed bottle was teeming with small living things that he could observe more clearly with the newly invented microscope. The sealed bottle showed no signs of life. This test certainly excluded spontaneous generation as a viable theory, except, as scientists of the day noted, that Spallanzani had deprived the closed bottle of air; air was thought a necessary ingredient for spontaneous generation. So although his experiment succeeded, a strong rebuttal blunted his claims.[507]

3a. Darwin's View on Spontaneous Generation

Pasteur's experiments closed the door on spontaneous generation and Darwin was aware of that fact. Yet, Darwin believed that a naturalistic cause for the creation of the first creatures on earth would one day be shown and in the third edition of the *Origin of the Species* (1861), page 514, he writes:

It is no valid objection that science as yet throws no light on the far higher problem of the essence or origin of life.

He takes an entirely different view in the second edition of *Origin of Species* (1860), which remains the same in the third through sixth editions, where he uses the "Creator", not science, as the cause of first life. He writes:

> There is grandeur in this view of life, with its several powers, having been originally breathed by the Creator into a few forms or into one. [508]

When his first edition is examined (1859), we again find a slightly different view where he did not use the words "by the Creator", instead, he implies a Creator. Darwin writes:

> There is grandeur in this view of life, with its several powers, having been originally breathed into a few forms or into one.[509]

Darwin was criticized for his inconsistency of using special creation by specifically crediting the Creator as the cause of the first creatures. Darwin, like his grandfather who was the first Englishman to write about evolution, was not being consistent in that he inserted special creation as the cause of the first life forms and then nature as the cause for the following life forms, including man.

Darwin privately retracted his claim of a Creator being involved with creation of the first life forms in his March 29, 1863 letter to his friend, English botanist and explorer Joseph D. Hooker, where Darwin writes:

> But I have long regretted that I truckled to public opinion and used Pentateuchal [relating to the first five books of the Jewish or Christian Scriptures] term of creation, by which I really meant "appeared" by some wholly unknown process.[510]

It is difficult to understand Darwin's regret of using Pentateuchal creation by God as opposed to an "unknown creation process," that was believed to be naturalistic, yet that was Darwin's written regret. The 1863 letter to Hooker was written before his release of the fourth (1866), fifth (1869), and sixth (1872) editions of the *Origin of Species* in which he continued to use the "Creator" as the cause of the first life forms. Why did he "truckle" to the public in the following editions of his book by using a "Creator" and state his regrets for doing so to his friend Hooker?

In an 1871 letter to Hooker, just before his sixth and last edition of the *Origin of Species* (1872), Darwin again writes his view about spontaneous generation as follows:

> It is often said that all the conditions for the first production of a living being are now present, which could ever have been present. But if (and oh what a big if) we could conceive in some warm little pond with all sort of ammonia and phosphoric salts,—light, heat, electricity present, that a protein compound was chemically formed, ready to undergo still more complex changes, at the present such matter would be instantly devoured, or absorbed, which would not have been the case before living creatures were formed.[511]

This was a very naturalistic view. Darwin's mixed use of the Creator and naturalistic approaches spans his six editions over thirteen years. He had multiple opportunities to change his public creation claim from the Creator to nature and natural selection, but never did in the *Origin of Species*. He had one side of himself shown to his friend Hooker and another side to the public, which leaves questions as to his motivations for taking that approach. Perhaps Darwin's active support of Huxley's *atheistic* X-Club gives a hint and reveals some of Darwin's private side. Perhaps his written opposition to special creation in his 1871 book, *Descent of Man*, offers another hint of his private side. Could his use of a "public-private" dichotomy about the creation of the first living creatures be a public relations means of having his ideas more widely accepted by the use of the term "Creator," thereby softening or masking the public's view of natural selection being atheistic? His private naturalistic view has all the earmarks of being his true view. That would mean the first life forms were created by a naturalistic model, at least for Darwin. His writing shows a "no special creation" allowed and "naturalistic only" worldview, which may reveal his core belief from which his views emanated, including a "life from non-life" belief called spontaneous generation, also called archebiosis. This view was not science in Darwin's time and is not science today. It is a religious view. The only cause and effect model which addresses spontaneous generation, is independently repeatable, is rooted in nature, and may be called science, is Pasteur's, which is discussed next.

3b. Pasteur's Spontaneous Generation Model

In Louis Pasteur's time (1822–1895), which was also Gregor Mendel's time, which was also Darwin's time, it was commonly accepted that life arose from nonlife through spontaneous generation. Then Pasteur's experiments showed it did not. In the 1860s, after Darwin's publication of several editions of *Origin of Species*, spontaneous generation was still a subject of debate in the exalted French Academy of Sciences.[512] Pasteur proved his "cause and effect" view to be correct with experiments that were independently repeatable. Though it may be argued that his experiment was not all-encompassing to cover all possible alternatives over all time, Pasteur's experiment shut the door on spontaneous generation as a commonly accepted model of creating life from nonlife: the cause and effect of the creation of the first life on earth was transformed into a faith system.

Pasteur's experimental design made use of a swan-neck flask into which fermentable juice was placed in a flask, and after sterilization, the neck was heated and drawn out as a thin tube taking a gentle downward and then upward arc, resembling the neck of a swan. The end of the neck was then sealed. As long as it was sealed, the contents remained unchanged. If the flask was opened by nipping off the end of the neck, air entered, but dust was trapped on the wet walls of the neck. Under this condition, the fluid would remain forever sterile, showing that air alone could not trigger growth of microorganisms. If, however, the flask was tipped to allow the sterile liquid to touch the contaminated walls, and this liquid was then returned to the broth, growth of microorganisms immediately began. In the words of Pasteur: "Never will the doctrine of spontaneous generation recover from the mortal blow of this simple experiment. No, there is now no circumstance known in which it can be affirmed that microscopic beings came into the world without germs, without parents similar to themselves."[513] No one since Pasteur has shown independently repeatable experiments that proved spontaneous generation takes place.

4. Darwinist Worldview: No Special Creation Allowed

The Darwinian worldview includes one core idea: there is no special creation. This core view of Darwinism is not science. It is a religious tenet. Without the core Darwinian component of "no special creation," all the other parts of the creation model would disintegrate, for they are but branches on that one tree. Disharmony about natural selection was set aside because of the intense focus against "special creation"; it was religion against religion, not science against religion. In one case, it pitted "no special creation" of Darwinism against Genesis. The Darwinian worldview allowed many different people, with different and conflicting views, to join under its one view of "no special creation," essentially for the purpose of eliminating God's direct involvement with creation and even eliminating God from the public forum and the primary and secondary educational environments.

This one view provided the common interest in those working together against religion. This one view was the motivation for Huxley forming the X-Club with its atheist members. Darwin was an ancillary member who approvingly and actively supported it. Darwinists did not all believe in natural selection, but they all believed in evolution by means other than special creation. Spencer, for example, was a Lamarckian. Wallace believed in natural selection, but not for the creation of man's "finer qualities." Huxley was a saltationist and did not accept natural selection. Lyell was a Lamarckian and never accepted natural selection. Asa Gray believed God directed variations in natural selection. For their own reasons, they each opposed special creation, which is special creation directly by God using saltation.

Some Darwinists accepted natural selection. Others rejected it. But all held that no creatures remained unchanged. The one that differed from this point was A. R. Wallace, the co-discoverer of natural selection, who thought that God created man's intelligence and "higher qualities," which he discussed in his 1869 article, "The Limits of Natural Selection as Applied to Man."[514] Wallace's many tests in that article proved that natural selection could not have created man's higher qualities. Darwin disagreed, but he never answered Wallace's tests. Darwinist Ernst Mayr writes:

> In the Darwinist worldview, the important point is whether evolution [by natural selection] is a natural phenomenon or something controlled by God.[515]

Ernst's statement is religious, not scientific. It is a fact that the two causes of creation—God and natural selection—cannot live together. This combining of the two is called theistic evolution, discussed in Chapter 4: Theistic Evolution. In other words, Ernst Mayr correctly believed that you must choose God or natural selection. This is a strange statement for a model that is said to be science. It is not claimed for models by Newton, Copernicus, Galileo, Kepler, Pasteur, Mendel, or any of the thousands of successful models of science. It is one more illustration that science does not conflict with religion, but testimonial models do. For most Darwinists, it is natural selection or God. Nowhere in science or in nature does one encounter such models, but one finds them in faith systems such as Darwinism. Choosing a religion or switching religions does take place, often with personal anguish. But the question of God never enters into models of nature that where nature links physical cause and effect. The model of "cause of creation and new creatures" is linked by testimony.

Newton believed in God, as did Copernicus, Pasteur, Bernoulli, and Mendel. Belief in God does not exclude observing nature and its rules of operation and developing successful models. Many scientists held a belief in the biblical God and in the models and rules of nature's operations. Darwinist proponents of natural selection must choose between God and nature because Darwinism is a faith system, and natural selection does not have the characteristics of science. Mayr's above quote demonstrates that the public debate about evolution consists of faith against faith or, said differently, Darwinism against all competing faith systems, especially that of special creation.

Worldview Summary

History has shown that multiple worldviews are typically needed to solve problems found in nature's physical world. Believing that only one worldview could be used to solve all problems has a history of omitting views necessary to solve problems. For example, those who held the mechanical worldview believed it could successfully be used to create models of nature in all subjects, such as biology, fields (electromagnetic waves), and chemistry. It was believed, for example, that biology would sooner or later be reduced to the laws of mechanics,[516] which was a belief, not science—a false belief.

History has shown that one worldview may not be likely to support the models of nature developed for problems outside that view, such as relativity models that could not be developed from an absolute worldview that was held by Newton. One worldview operating today, from which natural selection is formed, is that models of nature may be created using terms that are independent of nature's physical world and be science: but only faith-based or religious models could be created using such "independent of nature" terms.

Another worldview held by supporters of natural selection posits that denying God's creation of new creatures (a religion) proves that natural selection is correct, when in fact one has no relation to the other; thus, arguing against a religion does not prove natural selection to be correct. Brilliant men in one worldview became closed-minded and intolerant in another. Their inflexibility about their worldview biased them and limited their conceptualization abilities, thereby obscuring their intellectual capabilities. Such was the case with the great French philosopher Rene Descartes and his whirlpool (i.e., vortex) view of gravity. He proposed a nonmathematical model and suggested that the universe consists of huge whirlpools ("vortices") of cosmic matter. Our solar system would be only one of many such whirlpools.

This "one size fits all" worldview showed that holding to one view, with no others thought possible, could seduce one into thinking that his or her worldview applied to all problems and solutions involving nature, known and unknown. To hold to only one worldview and allow it to become the walls surrounding a "way to think" about nature makes the person a prisoner to that one view, which then forms the prison cell bars surrounding his or her thinking. Strangely, this prison is often not seen as restricting but rather as enlightening.

For a very long time, the premier worldview of modern science was the mechanical worldview. Then Copernicus introduced the sun-centered solar system as a new worldview. Then Newton introduced gravity (action at a distance) and the laws of motion, with the absolute concepts of space (infinite), time (unchanging), and velocity (unlimited). Then "fields" were introduced to form another worldview in which the mechanical worldview did not apply. Then the relativity worldview was introduced in which previous worldviews did not apply. Pasteur fully introduced the "germs" worldview; Mendel introduced the modern genetic worldview. These worldviews had one thing in common: all were used to create models of how the physical world of nature could be used to foretell what would take place in nature, followed by confirming observations in nature. All of them were used to create technologies and even industries that benefited men of all religions and views of nature.

Darwin's worldview was sufficiently simplistic and lacking of scientific merits that it allowed the attribution of facial expressions to "prove" evolution by natural selection, using the same test approach by which Wallace "proved" that some parts of the human body, such as the voice or brain, could not be created by natural selection. With an accommodating worldview, such as natural selection, a person can make observations and filter them through any physically undefined creation process to accommodate its tenets. For Darwin and his worldview, even facial expressions are capable of showing the common ancestry found in evolution by natural selection—this despite the fact that the words "facial" and "expressions" are not found in natural selection, and neither are the geometric configurations used to compose anything akin to facial expressions. Darwin wrote about facial expressions in his 1872 book, *The Expression of the Emotions in Man and Animals*. Darwin thought that facial expressions proved natural selection. Historian Janet Browne writes:

> He [Darwin] felt sure that some human [facial] expressions were universal, indicating mankind's single origin [common ancestor]. Moreover, he thought that the majority of human expressions were also identifiably the same as animal expressions, or at least their origins could be traced in animal movements and emotions, another sign of evolutionary connections [in his mind]. The human grimace of pain would be the same all over the globe, he imagined. It was equally recognizable in dogs. Surely he asked himself, this revealed the "mental continuities" between animals and humans?[517]

As shown in Browne's above quote, for Darwin, the *test* of natural selection and gradual creation was a person's smiling, or wincing, or frowning, or squinting, which, for him, proved that creatures changed slowly, little by little, into other creations; these expressions were sufficient proof, he thought, that evolution by natural selection took place: cause and effect were thus linked by testimony and faith. Actually, natural selection contains nothing about facial expressions; his worldview acted as evolutionary-colored glasses that revealed those "facts." The early use of photographs as a means of "scientifically" showing "facts" appears in *The Expression of the Emotions in Man and Animals*.[518] In contrast to Darwin was Sir Charles Bell, a knighted professor of surgery and a medalist who authored the book *The Hand: Its Mechanism and Vital Endowments as Evincing Design*. He also coauthored and published, with his brother, the two illustrated volumes of *A System of Dissection Explaining the Anatomy of the Human Body*. Bell's book shows his worldview that God created man. One need not agree with Bell's conclusions to recognize his high degree of medical and physiological knowledge and capabilities.

In 1829 Bell received the first medal awarded by the Royal Society—50 guineas; he was knighted by King William IV (1765–1837) in 1831.[519] Sir Charles Bell argues that muscles existed in the human face that were without analogy in lower animals,[520] meaning that no common ancestor could be involved—hence no evolution by natural selection. Different men, using different worldviews, conceive of models of creation that agree with their worldviews. No testimonial models possess the characteristics of the models of science, for science characteristics do not allow personal worldviews or models that allow them, such as natural selection.

EPILOGUE

A theory is an idea. The right idea can change the world by changing how people think and act. A theory may make a person famous, depending on the idea; it may even make him or her infamous, as has become the case for Darwin. Theories are typically causal, operating either in nature, as with the model of gravity, or within the mind, such as with belief systems like those surrounding natural selection. That is, theories express the cause of something, such as the creation of new creatures. Darwin proposed a causal theory: evolution through natural selection. He did not claim that evolution was natural selection or that it was a process, which would have been an error that one encounters today. Darwin claimed his theory as science, but he was wrong. Natural selection contains no parts of nature; any theory that purports to show that it creates new creatures must show how it operates in nature using nature's physical properties. That is not the case with natural selection. Darwin used arguments in place of nature—a practice that exists to this day.

Revealed in many scientific cause and effect models of nature, such as Pasteur's or Newton's, is how cause relates to effect by nature linking the two with nature's components. The rules that govern nature's links, operational processes, and the relationship of the model's many parts to each other and to the effects *are all contained in the model*, not in any testimony about the model or attributions of effect to it. In that revelation, given by the review of all successful causal, foretelling models of nature, we see a stream of common characteristics that show what science is and how it operates: those models' characteristics define science. For instance, Pasteur's vaccination model shows only nature's parts operating in a cause and effect manner, the effect being a cure and prevention of deadly diseases. Pasteur became the most famous scientist of his day because of his vaccination model that cured people. In another case, Archimedes became famous because of his buoyancy model, which showed the upward force exerted on a body in a fluid, like a hot-air balloon or a boat in water. Mendel became famous and known as the father of modern genetics because of his causal genetics model.

Causal models may also be formed without parts of nature; many such models exist, such as those in all of evolutionary biology and the major religions. Miracles are causal models. These models are empty of nature; they operate solely by testimonial attributions of effects that already exist to a causal model that cannot be shown to operate or exist in nature: they operate by inferences. Evolutionary models may operate by calling their inferences about the past "history," which inferences are not. Of those theories that are shown to physically exist in nature, examples of their operating parts of nature are time, distance, weight, force, germs, and electrical signals. The links between nature's many parts in a model and the effects that are created must be shown in the model. The body's major systems form components of causal models, such as skeletal system (with each bone being unique in shape and position), control-feedback systems, identification systems, reproductive systems, digestive systems, transportation systems, communications systems, the thousands of enzymes in the body, and so on. Each one is physical and observable. With those foretelling models that successfully operate in nature, cause and effect takes place without testimony. Science begins and ends with these models.

A causal model that has no physical components and is said to operate in nature is a spirit. Evolutionary biology is filled with such models that are causally empty of nature, like those of Buffon, Lamarck, Erasmus Darwin, William Charles Wells, Patrick Matthews, Edward Blyth, Alfred Russel Wallace, Charles Darwin, and all others who embrace models of creation, including special creation, use and disuse, arms races, adaptive landscapes, selection pressures, and so on. These "cause and effect" models have operations, relationships, and rules that do not exist in nature.

The NAS booklet, *A View*,[521] used evolution without a cause mentioned, which for the NAS is understood as being natural selection. The NAS incorrectly equates evolution and natural selection thus:

The concept of biological evolution [by natural selection] is one of the most important ideas ever generated by the application of scientific methods to the natural world. The evolution of all the organisms that live on Earth today from ancestors that lived in the past is at the core of genetics, biochemistry, neurobiology, physiology, ecology, and other biological disciplines. It helps to explain the emergence of new infectious diseases, the development of antibiotic resistance in bacteria, the agricultural relationships among wild and domestic plants and animals, the composition of Earth's atmosphere, the molecular machinery of the cell, the similarities between human beings and other primates, and countless other features of the biological and physical world. As Theodosius Dobzhansky wrote in 1973, in his article, "Nothing in biology makes sense except in the light of evolution."[522]

The claim by the NAS is entirely incorrect—no one needs or uses biological evolution in his or her work with nature. The characteristics of natural selection do not allow it to operate as a model in nature or as a causal model that is science. Pasteur did not use natural selection for infectious diseases. Mendel did not use natural selection for genetics. Newton did not need or use evolution or natural selection for his theories. Edward Jenner discovered a cure for smallpox without evolution or natural selection. William Harvey discovered the circulatory system with no need of Darwin's natural selection. Polio was conquered without natural selection or evolution. No one used natural selection other than for politically correct, inconsequential genuflections. The NAS statement that evolution lays at the core of genetics, biochemistry, and other biological disciplines[523] is empty of nature and empty of science. There is no biology in natural selection. One could go through four years of engineering, which is one of the most intense exposures to science, and then through graduate school to the doctorate and post doctorate level and never encounter evolution, natural selection, or Darwin. One could major in biomedical engineering and never encounter evolution, natural selection, or

Darwin for he had nothing to offer as a causal model that operated in nature. He is not needed nor used.

The NAS claim about evolution is akin to the philosopher's stone of alchemy: imagined, but never shown to exist or operate. It represents a "way of thinking" about nature not shown or observed in nature; that way of thinking is intended to support a worldview in opposition to opposing worldviews; a worldview that is not science but rather religion. The public debate surrounding natural selection has never has been "science against religion," even though it is often portrayed that way and sometimes, surprisingly, accepted by its opponents. The debate has always been about "religion against religion": the religion of Darwinism against other competing religions, mostly (in the literature) the religion of Darwinism against the religion of Christianity. The public debate has always been fought to convince others to accept one way of thinking about nature: the Darwinist way. At this time, and likely forever, the actual model of creation, the model of cause and effect in the physical world of nature, is still a mystery. The responsible creation model is an unknown and likely will be forever; that is the correct way to think about evolution by any model because it is accurate.

R. A. Fisher's words, shown at the very beginning of chapter 1, are repeated here:

> Natural Selection is not Evolution. Yet, ever since the two words have been in common use, the theory of Natural Selection has been employed as a convenient abbreviation for the theory of Evolution by means of Natural Selection, put forward by Darwin and Wallace. This has had the unfortunate consequence that the theory of Natural Selection itself has scarcely ever, if ever, received *separate consideration*. [Italics added][524]

This *separate consideration* of natural selection is shown in this book by natural selection being examined from several perspectives. Natural selection is shown to fail as a creation model that is meant to show "evolution by natural selection" in nature. Natural selection can never be a creation model or a cause of evolution because every term in the model, along with the model, is independent of nature: they possess no physical characteristics or existence. Natural selection does not exist in the physical world, but only in the world of inference, logic, and imagination under the umbrella of the Darwinist worldview: none of this is causal in nature, but may be quite causal within one's mind as a way of thinking. Natural selection is unsupported by observations that stem from the model itself. Observations, when made, are by testifiers attributing observations to natural selection. The need for attribution shows natural selection to be unscientific and a failure at creating new creatures or adaptations. Further, there are no *incomplete* intermediates which are mandated by any gradual creation model such as natural selection. No scientific or other method can resurrect natural selection from its inoperative state in nature and science.

The separate consideration given here shows that natural selection is the object of a religion called Darwinism whose rallying point and major tenet is "no special creation"—hardly an undertaking of scientific pursuit. It even fails in that attempt. Railing against a religion, such as Christianity, is not a scientific undertaking or one that shows natural selection to have any merit other than serving as a testimonial platform for the competing religion of Darwinism. Darwinism and natural selection survive by testimony, inferences, debating tactics, wordsmithing, logic, and rhetorical conflation, but not by nature's independently repeatable physical operations of cause and effect. Despite the constant watering of natural selection with testimony, attribution, and politicization, the model has already withered on the vine to an intellectually lifeless state. Natural selection is destined to fade into a curiosity of history—along with the testimonial discipline of evolutionary biology, the unscientific arguments of Darwin, and the religion of Darwinism—where natural selection will be discussed alongside phrenology, phlogiston, and the philosopher's stone. What is left to be done is to remove the religion of Darwinism from the science and biology classrooms and textbooks along with its major tenet: *evolution through natural selection.*

Notes and References

[1] From the Danbury Baptist Association, https://jeffersonpapers.princeton.edu/selected-documents/danburybaptist-association, 30 Dec 2014.

[2] Ibid.

[3] Ibid.

[4] R. A. Fisher, Sc.D., F.R.S., *The Genetical Theory of Natural Selection* (Oxford, UK: Clarendon Press, 1930), opening lines, vii.

[5] Anders Hald (1913–2007) was a Danish statistician, writer of the history of statistics, and professor at the University of Copenhagen from 1960 to 1982.

[6] "Genetical Theory of Natural Selection," http://en.wikipedia.org/wiki/The_Genetical_Theory_of_Natural_Selection, 24 Jan 2014.

[7] John Gribbin, *Science, A History 1543–2001* (New York: BCA, 2001), 333.

[8] Ibid.

[9] Ernst Mayr, *What Evolution Is* (New York: Basic Books, 2001), 80.

[10] Sherri L. DeFauw, "Evolution: The Highlights," *Humanism Today* 39 (2014), http://www.docstoc.com/docs/27901817/EVOLUTION-THE-HIGHLIGHTS, 9 June 2014

[11] Richard Milner, *Encyclopedia of Evolution, Humanity's Search for Its Origins*," foreword by Stephen Jay Gould, An Owl Book (New York: Henry Holt and Company, 1990): 159.

[12] "Orthogenesis," 14 April 2010, http://en.wikipedia.org/wiki/*Genesis*.

[13] Peter J. Bowler, "The Changing Meaning of Evolution," *Journal of the History of Ideas* 36, no. 1 (Jan.– Mar. 1975): 95–114.

[14] Herbert Spencer, *Social Statics: or, The Conditions essential to Happiness specified, and the First of them Developed* (London: John Chapman, 1851), 127.

[15] Ibid., 395.

[16] "Transmutation of Species," http://en.wikipedia.org/wiki/Transmutation_of_species, 29 July 2009.

[17] The Descent of Man, Vol. 1, 1871, 1st Edition, Murray Pub: John Murray, Albemarle Street, London, 1871, http://darwin-online.org.uk/EditorialIntroductions/Freeman_TheDescentof Man.html, January 20, 2015.

[18] Charles Darwin, *On the Origin of Species by Means of Natural Selection, or the Preservation of Favoured Races in the Struggle for Life*, 6th ed. (London, UK: John Murray, 1872), 282.

[19] Charles Darwin, *On the Origin of Species by Means of Natural Selection, or the Preservation of Favoured Races in the Struggle for Life*, 6th ed. (London, UK: John Murray, 1872), 282.

[20] Darwin, *Descent of Man*, 343, 459.

[21] Darwin, *Origin of Species*, 6th ed., 201.

[22] Francis Darwin, ed., "Life and Letters of Charles Darwin," vol. I (London, UK: John Murray, 1888), 437.

[23] "Saltation," http://www.oxforddictionaries.com/us/definition/american_english/saltation, 24 April 2014.

[24] Niles Eldredge and Stephen Jay Gould, "Punctuated Equilibrium, an Alternative to Phyletic Gradualism," in *Models in Paleobiology*, ed. Schopf (San Francisco: TJM Freeman, Cooper & Co, 1972), 82– 115, 84.

[25] "Stasis in The Fossil Records," http://www.living-fossils.com/2_1.php, 28 April 2014.

[26] Samuel Butler, *Evolution, Old and New or, the Theories of Buffon, Dr. Erasmus Darwin and Lamarck, as compared with that of Charles Darwin*, 1879, (New York, E. P. Dutton &Company, 681 Fifth Avenue: 1882), 166-167.

[27] Ibid.

[28] "Creationism," Stanford, http://plato.stanford.edu/entries/creationism/, 6 May 2014.

[29] Ibid.

[30] Darwin, *Origin of Species* (1872), 282.

[31] Darwin, *Origin of Species*, (1859), 343, 459.

[32] George Mivart, *On the Genesis of Species* (New York: D. Appleton & Co., 1871), 45.

[33] Ibid., 87.

[34] "Abio*genesis*," http://en.wikipedia.org/wiki/Abio*Genesis*, 22 April 2012.

[35] Stephen Jay Gould, *Wonderful Life: The Burgess Shale and the Nature of History* (New York: W. W. Norton & Company, 1989,) 60.

[36] Prokaryote, http://autocww2.colorado.edu/~toldy2/E64ContentFiles/VirusesMoneransAnd Protists/Prokaryote.html

[37] "File: Average prokaryote cell—en.svg," http://en.wikipedia.org/wiki/Image:Average_prokaryote_cell_en.svg, 25 Oct 08.

[38] Ibid.

[39] Stephen Jay Gould, *Wonderful Life: The Burgess Shale and the Nature of History* (New York: W. W. Norton & Company, 1989,) 58.

[40] Ibid.

[41] Ernst Mayr, *What Evolution Is* (New York: Basic Books, 2001), 228.

[42] Stephen Jay Gould, *Wonderful Life: The Burgess Shale and the Nature of History* (New York: W. W. Norton & Company, 1989,) 58.

[43] Ernst Mayr, *What Evolution Is* (New York: Basic Books, 2001), 228.

[44] "The Evolution of Life on Earth," *Scientific American* (October 1994). http://brembs.net/gould.html, 14 Jan 2012.

[45] Stephen Jay Gould, *Wonderful Life: The Burgess Shale and the Nature of History* (New York: W. W. Norton & Company, 1989,) 58, 59.

[46] The creatures were intermediate by classification based on shapes, not by biological ancestry.

[47] Milner, *"Encyclopedia of Evolution,"* 158.

[48] Ibid.

[49] Ibid.

[50] Ibid.

[51] Darwin, *Origin of Species* (1859), 171.

[52] Darwin, *Origin of Species* (1872), 134.

[53] Darwin, *Origin of Species* (1859), 343, 459.

[54] Ibid., 280.

[55] Ibid., 302.

[56] Ibid.

[57] Ibid.

[58] Ernst Mayr, *What Evolution Is* (New York: Basic Books, 2001), 275.

[59] Milner, *"Encyclopedia of Evolution,"* 157.

[60] *Science and Creationism: A View from the National Academy of Sciences*, 2nd ed., Steering Committee on Science and Creationism, National Academy of Sciences (1999); ISBN: 0-309-53224-8. A free copy may be downloaded from: http://www.nap.edu/catalog/6024.html.

[61] Ernst Mayr, *What Evolution Is* (New York: Basic Books, 2001), 275.

62 Ibid., 276.

63 Ibid., 276.

64 *Science and Creationism.*

65 "What Is Evolution?" *BioLogos* (2 August 2013), http://biologos.org.

66 *Science and Creationism.*

67 Samuel Butler, *Evolution, Old and New or, the Theories of Buffon, Dr. Erasmus Darwin and Lamarck, as compared with that of Charles Darwin,* 1879, (New York, E. P. Dutton &Company, 681 Fifth Avenue: 1882),, 355-356

68 Samuel Butler, *Evolution, Old and New or, the Theories of Buffon, Dr. Erasmus Darwin and Lamarck, as compared with that of Charles Darwin,* 1879, (New York, E. P. Dutton &Company, 681 Fifth Avenue: 1882), 356.

69 Ibid.

70 Janet Browne, *The Power of Place* (New York: Alfred A. Knopf, 2002), 55.

71 Ibid., 55–56.

72 Ibid., 55–56.

73 Ibid., 55–56.

74 Darwin, *Origin of Species* (1859), 478.

75 Lightman, Bernard. "On Tyndall's Belfast Address, 1874." *BRANCH: Britain, Representation and Nineteenth-Century History.* ed. Dino Franco Felluga. Extension of *Romanticism and Victorianism on the Net.* Web: http://www.branchcollective.org/?ps_articles=bernard-light-man-on-tyndalls-belfast-address1874 [January 20, 2015].

76 John Tyndall, "Address Delivered before the British Association Assembled at Belfast, With Additions, 1874," http://www.victorianweb.org/science/science_texts/belfast.html, 7 Jan 2014.

77 Darwin, *Origin of Species* (1859). Darwin argues against God (special creation) at many points in his books. Examples of his rejection of God as creator of new creatures can be seen on pages 55, 167, 413.

78 Jonathan Weiner, *The Beak of the Finch, A Story of Evolution in Our Time* (New York: Alfred A. Knopf, 1995), 6.

79 Herbert Wendt, *In Search Of Adam* (Boston, MA: Houghton, Mifflin Company, Riverside Press, 1956), 89.

80 *The Free Dictionary.com,* 24 Sep 2004.

81 Milner, *"Encyclopedia of Evolution,* 371.

[82] Richard Goldschmidt, *Material Basis of Evolution*, with an introduction by Stephen Jay Gould, 1940, 1982, New Haven and London, Yale University Press, xl.

[83] Herbert Wendt, *In Search Of Adam* (Boston, MA: Houghton, Mifflin Company, Riverside Press, 1956), 89.

[84] Ibid., 89.

[85] Ibid., 87.

[86] Ibid., 87, 89.

[87] Vernon Lyman Kellogg, *Darwinism To-Day* (New York: Henry Holt and Company, 1907), 33, quoted in "Chance Variation and Evolutionary Contingency: Darwin, Simpson (The Simpsons), and Gould," John Beatty, Department of Philosophy; University of British Columbia, Vancouver, BC V6T 1Z1, Canada; john.beatty@ubc.ca.

[88] Gordon Rattray Taylor, *The Great Evolution Mystery* (New York: Harper & Row, 1983), 116.

[89] Ibid.

[90] Darwin, *Origin of Species* (1859). For "special creation," see 482, 488. For "independent creation," see 139, 355, 489, 398, 406, and 435.

[91] "Special Creation," http://www.blueletterbible.org/faq/don_stewart/don_stewart_615.cfm, 29 April 2014.

[92] Ibid.

[93] Ibid.

[94] Janet Browne, *Power of Place*, 319.

[95] Ibid.

[96] Alfred Russel Wallace, *Darwinism, An Exposition of The Theory of Natural Selection with Some of Its Applications* (London and New York: Macmillan And Co., 1889), 427.

[97] "Punctuated equilibrium", http://en.wikipedia.org/wiki/Punctuated_equilibrium, 7 June 2014.

[98] "Phyletic gradualism," http://en.wikipedia.org/wiki/Phyletic_gradualism, 17 Nov 2011.

[99] Mivart, *Genesis of Species*, (New York: D. Appleton And Company, 549 & 551 Broadway, 1871), 141.

[100] Ibid., 111.

[101] Ibid.

[102] "Darwinism: Being an Examination of Mr. St. George Mivart's 'Genesis of Species,'" Chauncey Wright, http://darwin-online.org.uk/converted/Ancillary/1871_Wright_A576.html, 1 April 2014.

103 Mivart, *Genesis of Species*, 141.

104 Ibid., 127.

105 Ibid., 111.

106 Francis Galton, *Hereditary Genius; An Inquiry into its Laws and Consequences* (London and New York: MacMillan and Co., 1892), 368.

107 Ibid., 369.

108 Ibid., 369.

109 Ernst Mayr, *What Evolution Is* (New York: Basic Books, 2001), 276.

110 James D. Watson, *The Double Helix: A Personal Account of the Discovery of the Structure of DNA* (New York: Mentor Books, 1969), 53.

111 Hiram Caton, "Getting Our History Right: Six Errors about Darwin and His Influence," http://www.darwin-legend.org/html/Charles-Darwin-Six-Errors.htm, 27 Aug 08.

112 Darwin, *Origin of Species* (1859), 355.

113 Alpheus S. Packard, *Lamarck, Founder of Evolution*, 94. Packard's quote cites *Origin of Species*, 3rd edition, 1861, Historical Sketch, xiv.

114 Ibid., 94.

115 Arnold C. Brackman, *A Delicate Arrangement, The Strange Case of Charles Darwin and Alfred Russel Wallace* (New York: Times Books, 1980), inside front flap.

116 Ibid.

117 Alfred Russel Wallace, "On the Law Which Has Regulated the Introduction of New Species (S20: 1855) (aka: Sarawak law)," http://people.wku.edu/charles.smith/wallace/S020.htm, 16 Oct 2012.

118 Ibid.

119 David Stove, *Darwinian Fairytales* (New York: Encounter Books, 1995), 28.

120 Janet Browne, *Power of Place*, 12.

121 Ibid.

122 David N. Stamos, *Darwin and the Nature of Species* (State University of New York Press), 161.

123 "Vestiges of the Natural History of Creation," http://en.wikipedia.org/wiki/Vestiges_of_the_Natural_History_of_Creation, 14 Dec 09.

124 Gregory Tate, the *Poet's Mind*, The Psychology of Victorian Poetry 1830-1870 (Oxford, United Kingdom, Oxford University Press, 2012), 99.

125 Ibid.

126 "Disraeli, Benjamin," http://en.wikipedia.org/wiki/Benjamin_Disraeli, 21 Nov 2013.

127 Ibid.

128 Ibid.

129 Ibid.

130 Charles Darwin, *Origin of Species by Means of Natural Selection, or the Preservation of Favoured Races in the Struggle for Life*, 3rd ed. (London, UK: John Murray, 1861), xv–xvi.

131 A. De Vries, "The Enigma of Darwin," *Clio Medica* 19, no. 1–2 (1984):136–155, 145 (quoted in *Evolutionary Naturalism: An Ancient Idea*, http://www.answersinGenesis.org/articles/tj/v15/n2/naturalism, 21 Nov 2013).

132 Pearson, Paul N, reprint of: James Hutton, *An Investigation of the Principles of Knowledge and of the Progress of Reason, from Sense to Science and Philosophy,* (Selected extracts are reproduced by permission of the Royal Society of Edinburgh from *Elements of Agriculture* by James Hutton, an unpublished manuscript (1794–1797) held on deposit at the National Library of Scotland), 500501.

133 Arnold C. Brackman, *A Delicate Arrangement, The Strange Case of Charles Darwin and Alfred Russel Wallace* (New York: Times Books, 1980), 75.

134 C. R. P. George, "William Charles Wells (1757-1815)—a nephrologist of the Scottish enlightenment," Nephrol Dial Transplant (1996) 11: 2513-2517, 1996 European Renal Association-European Dialysis and Transplant Association.

135 Charles Darwin, *On the Origin of Species by Means of Natural Selection, or the Preservation of Favoured Races in the Struggle for Life*, 4th ed. (London, UK: John Murray, 1866), Historical Sketch, xiv-xv.

136 Paul N. Pearson, "In Retrospect," *Nature* v. 425 #6959, 665. Comments on Hutton's 3-volume, 1794 work, *An Investigation of the Principles of Knowledge and of the Progress of Reason, from Sense to Science and Philosophy,* http://thedispersalofdarwin.blogspot.com/2007/06/more-on-james-hutton.html, 16 October 2003.

137 Charles Hodge, *What Is Darwinism?* (Princeton, NJ, New York: Scribner, Armstrong, And Company, 1874), 51.

138 Ibid.

139 Samuel Butler, *Evolution, Old and New,* (New York, E. P. Dutton &Company, 681 Fifth Avenue: 1882), 64.

140 Ibid., 29.

141 Wilberforce, Samuel (1860). "(Review of) 'On the origin of species'". *Quarterly Review*: 225–264.

[142] *Evolutionary Psychology* 5, no. 1: 52–69 Hiram Caton, Griffith University, Nathan 4111, Australia, hcaton2@bigpond.net.au, "Getting Our History Right," http://www.epjournal.net/articles/getting-ourhistory-right-six-errors-about-darwin-and-his-influence/, January 15, 2014.

[143] Ibid.

[144] Evolutionary Psychology 5, no. 1: 52–69 Hiram Caton, Griffith University, Nathan 4111, Australia, hcaton2@bigpond.net.au, "Getting Our History Right," http://www.epjournal.net/articles/getting-ourhistory-right-six-errors-about-darwin-and-his-influence/, 15 January 2014.

[145] Ibid.

[146] Darwin, *Origin of Species* (1866), xiv.

[147] Thomas Malthus, *An Essay on the Principle of Population, or a View of its Past and Present Effects on Human Happiness; with an Inquiry into our Prospects respecting the Future Removal or Mitigation of the Evils which it Occasions* (London: John Murray 1826).

[148] Wallace brought up this same fact about breeders' creatures reverting to their former states in the wild.

[149] "Edward Blyth," http://en.wikipedia.org/wiki/Edward_Blyth, 7 Jan 2014

[150] Ibid. Credit for the information presented on the two quotes go to: Andrew J. Bradbury, "Blyth, Edward - Did Darwin Plagiarize From Blyth," http://www.bradburyac.mistral.co.uk/dar7.html, 25 May 2011. More information on Blyth, Darwin, and priority is presented on: http://www.bradburyac.mistral.co.uk/dar0.html, 12 April 2014.

[151] Andrew J. Bradbury, "Blyth, Edward - Did Darwin Plagiarize from Blyth", available at http://www.bradburyac.mistral.co.uk/dar7.html, 25 May 2011.

[152] Wallace, "On the Law," http://people.wku.edu/charles.smith/wallace/S020.htm, 16 Oct 2012.

[153] Janet Browne, *Power of Place*, 32.

[154] Ibid., 14.

[155] Ibid., 32.

[156] Ibid., 14.

[157] Ibid., 32.

[158] Ibid., 39.

[159] Ibid., 34.

[160] Ibid., 37-38.

[161] Ibid., 37-39.

[162] "Darwin-Wallace Paper (complete)," http://www.indiana.edu/~koertge/H205c/, 26 Jan 2014

[163] Ibid., 35.

[164] Ibid., 35.

[165] Ibid., 35.

[166] Brackman, *Delicate Arrangement,* inside book jacket's front flap.

[167] Ibid., 62, 65.

[168] Ibid., 65.

[169] Ibid.

[170] Browne, *Power of Place,* 40.

[171] Ibid., 42.

[172] Brackman, *Delicate Arrangement,* 195.

[173] Ibid., inside book jacket's front flap.

[174] "Transmutation of species," http://en.wikipedia.org/wiki/Transmutation_of_species, 24 March 2012.

[175] Hiram Pendleton Caton III (16 August 1936 in Concord, North Carolina, USA–13 December 2010 in Ingham, Queensland) was a professor of politics and history at Griffith University, Brisbane, Australia, until his retirement. He was an ethicist, a Fellow of the Australian Institute of Biology (since 1994), and a founding member of the Association for Politics and the Life Sciences. He was an officer of the International Society for Human Ethology. Caton held a National Humanities Fellowship at the National Humanities Center in 1982–1983. He was the inaugural professor of humanities at Griffith University in Brisbane, and later the professor of politics and history and head of the School of Applied Ethics there. Source: Hiram Caton, http://en.wikipedia.org/wiki/Hiram_Caton, 13 March 2014.

[176] Evolutionary Psychology 5, no. 1: 52–69 Hiram Caton, Griffith University, Nathan 4111, Australia, hcaton2@bigpond.net.au, "Getting Our History Right," http://www.epjournal.net/articles/getting-ourhistory-right-six-errors-about-darwin-and-his-influence/, 15 January 2014.

[177] Ibid.

[178] "*Journal of the Proceedings of the Linnean Society of London. Zoology,*" 3 (20 August): pp. 45-50, http://darwin-online.org.uk/content/frameset?itemID=F350&viewtype=text&pageseq=1; 12 February 2015.

[179] Darwin, *Origin of Species* (1859), 490.

[180] Ibid., 84–85.

[181] P. Matthew, "Nature's Law of Selection," *Gardeners' Chronicle and Agricultural Gazette* (7 April 1860): 312–313.

[182] Hiram Caton, Review of Pietro Corsi's book, *The Age of Lamarck: Evolutionary Theories in France 1790–1830*.

[183] Ibid., 84–85.

[184] "Edward Blyth," http://en.wikipedia.org/wiki/Edward_Blyth, 23 Aug 2011.

[185] Edward Blyth, "An Attempt to Classify the 'Varieties' of Animals with Observations on the Marked Seasonal and Other Changes Which Naturally Take Place in Various British Species, and Which Do Not Constitute Varieties," *The Magazine of Natural History,* vol. 8, no. 1 (January 1835): 40–53.

[186] Alfred Russel Wallace, "On the Tendency of Varieties to Depart Indefinitely from the Original Type (S43: 1858)," http://people.wku.edu/charles.smith/wallace/S043.htm, [This is Wallace's Ternate Paper], 27 Jan 2015.

[187] Ibid.

[188] W. C. Wells, "Two Essays: Upon a Single Vision with Two Eyes, the On Dew and An Account of a Female of the White Race of Mankind, Part of Whose Skin Resembles That of a Negro, with Some Observations on the Cause of the Differences in Colour and Form between the White and Negro Races of Man." London, Archibald Constable and Co., Edinburgh, Longman, Hurst, Rees, Orme, and Brown, and Hurst, Robinson, And Co. London, 1818, 435-436.

[189] Brackman, *Delicate Arrangement,* 326n.

[190] Ibid., 19.

[191] "Theistic evolution," http://en.wikipedia.org/wiki/Theistic_evolution, 12 Feb 2013.

[192] BioLogos statement of belief: "*We believe* that the diversity and interrelation of all life on earth are best explained by the God-ordained process of evolution with common descent. Thus, evolution is not in opposition to God, but a means by which God providentially achieves his purposes. Therefore, we reject ideologies that claim that evolution is a purposeless process or that evolution replaces God."

[193] In this statement, the NAS shows evolution to be an "effect," something that is "caused or created." Darwin writes that it was caused by "natural selection." The NAS writes that evolution is driven by "various processes" that are "physical and biological" but does not mention natural selection.

[194] Science and Creationism. In this statement, the NAS shows evolution to be an "effect," something that is "caused or created." Darwin writes that it was caused by "natural selection." The NAS writes that evolution is driven by "various processes" that are "physical and biological" but does not mention natural selection.

[195] Science and Creationism.

[196] Science and Creationism.

[197] Browne, *Power of Place*, 310.

[198] Science and Creationism.

[199] Science and Creationism.

[200] Browne, *Power of Place*, 310.

[201] Ibid.

[202] "Religious," http://www.merriam-webster.com/dictionary/religious, 31 July 2009.

[203] "Religion," en.wikipedia.org/wiki/Religious.

[204] Robert Charles Winthrop, *Addresses and Speeches on Various Occasions*, vol. 1 (Boston: Little Brown, 1852), 172.

[205] Fr. Copleston vs. Bertrand Russell: The Famous 1948 BBC Radio Debate on the Existence of God [And The Moral Argument], http://www.biblicalcatholic.com/apologetics/p20.htm, February 6, 2015.

[206] Robert Charles Winthrop, *Addresses and Speeches on Various Occasions*, vol. 1 (Boston: Little Brown, 1852), 172.

[207] "95%+ of Our Founders Fathers were Bible Believing Christians!" http://jjusa.org/?p=1881, 5 May 2014.

[208] Ibid.

[209] Benjamin Franklin speech to the Constitutional Convention, June 28, 1787, Green Mountain Scribes, http://greenmountainscribes.wordpress.com/2011/11/20/benjamin-franklin-speech-to-the-constitutionalconvention-june-28-1787/, 2 May 2014.

[210] Ibid.

[211] Charles Darwin, *The Descent of Man*, Vol. 1, 1871, (Albemarle Street, London, Murray Pub: John Murray, 1871), 153.

[212] Darwin, *Origin of Species* (1859), 243–244.

[213] Ibid.

[214] Letter 2814, Darwin, C. R. to Gray, Asa, 22 May [1860], http://www.darwinproject.ac.uk/letter/entry-2814, June 2014: "[I] had no intention to write atheistically...I am inclined to

look at everything as resulting from designed laws, with the details, whether good or bad, left to the working out of what we may call chance."

[215] Alfred Russel Wallace, "The Limits of Natural Selection as Applied to Man (S165: 1869/1870)," http://www.wku.edu/~smithch/wallace/S165.htm, 9 Sep 09.

[216] Ibid.

[217] Browne, *Power of Place*, 319.

[218] Darwin, *Origin of Species* (1859), 459.

[219] Wallace's God is not the God of Genesis; it may be a pantheist God, an Intelligent Force, or some combination of characteristics that do not readily lend to identifying a particular God. See: Charles Hodge, *What Is Darwinism?* (Princeton, NJ, New York: Scribner, Armstrong, And Company), 1874. http://www.gutenberg.org/files/19192/19192-h/19192-h. htm#Relation_of_Darwinism_to_Religion, 30 Nov 2013, 11.

[220] *Science and Creationism*, 32.

[221] Ibid.

[222] Wallace, *Darwinism*, 113.

[223] Science and Creationism.

[224] Ibid., 31.

[225] It is impossible to determine how many "good" variations were accumulated to create each organ, body part, chemical, electrical signal, or system of organs. The variations are not visible, leave no residual marks, and are never observed in the fossils. They are also undefined.

[226] No variations, good or bad, can be moved or discarded once they are created. They are permanently fixed in a position in an accumulation, which is also fixed; neither is ever defined in nature. The only means of removing bad variations is to remove the entire creature. Darwin never discussed removal of variations in nature in any detail and always assumed that they were somehow removed without writing about how nature accomplished that feat. One good variation that is followed by one bad variation destroys all creation up to that point, forcing evolution by natural selection to start all over again. The end result is that creation by natural selection is, at most, an infinite series of creations and destructions with no creation of new body parts or new creatures ever taking place.

[227] "Selection" of a good variation, or any variation, cannot be performed until direction is set first. Only "direction" differentiates good and bad variations, which is toward the final goal of a new adaptation or new creature. This has the appearance of a back door through which a creator enters.

228 *Science and Creationism.*

229 Ernst Mayr, *What Evolution Is* (New York: Basic Books, 2001), 276.

230 Ibid.

231 Clarence Darrow, *The World's Most Famous Court Trial*, Tennessee Evolution Case, (Cincinnati, Ohio, National Book Company, 2010), 177.

232 Ibid.

233 Justice Paul E. Pfeifer, "Dec. 15, 1999, The Scopes Monkey Trial," https://www.supremecourt. ohio.gov/SCO/justices/pfeifer/column/1999/jp121599.asp, January 28, 2015.

234 Ibid.

235 Clarence Darrow, *The World's Most Famous Court Trial*, Tennessee Evolution Case, (Cincinnati, Ohio, National Book Company, 2010), 296-302. The discussion about the age of the earth spans several pages of the court records. There is a bantering about by the attorneys which is not necessary for the age of the earth and not included in the quotes given here.

236 Ronald L. Numbers, *Darwinism Comes to America* (Cambridge, MA: Harvard University Press, 1998), 82.

237 Theodosius Dobzhansky, http://en.wikipedia.org/wiki/Theodosius_Dobzhansky, 28 Nov 2013.

238 Theodosius Dobzhansky, "Nothing in biology makes sense except in the light of evolution," *The American Biology Teacher* (March 1973), http://www.pbs.org/wgbh/evolution/ library/10/2/text_pop/l_102_01.html, 5 Aug 09.

239 Ibid.

240 Ibid.

241 The four personifications are as follows: 1. Natural selection may cause a living species to respond to the challenge by adaptive genetic changes. 2. There is, of course, nothing conscious or intentional in the action of natural selection. 3. Natural selection is at one and the same time a blind and creative process. 4. Natural selection does not work according to a foreordained plan, and species are produced not because they are needed for some purpose but simply because there is an environmental opportunity and genetic wherewithal to make them possible. Personifications amount to deifications.

242 Ibid.

243 Dobzhansky, "Light of Evolution."

244 Ibid.

245 Ibid.

[246] Pierre Teilhard de Chardin, New World Encyclopedia, http://www.newworldencyclopedia.org/entry/Pierre_Teilhard_de_Chardin, 18 Jan 2015.

[247] Ibid. There is literally nothing that one can do in nature with such claims. The only alternative that is available is to accept or reject his view, much as one does with any claim that necessitates faith.

[248] "Light of the World," http://en.wikipedia.org/wiki/Light_of_the_World, 11 Nov 2012.

[249] American Standard Bible, John 8:12.

[250] Chanukkah, http://www.jewfaq.org/holiday7.htm, 10 March 2015

[251] Dobzhansky, "Light of Evolution."

[252] Ibid.

[253] Ibid.

[254] Darwin, *Origin of Species* (1859), 186.

[255] Quoted by Michael Denton in *Evolution: A Theory in Crisis* (Bethesda, MD: Adler & Adler, 1986), 217; original source: Dewar, D. *More Difficulties of the Evolution Theory* (London, UK: Thynne & Co, 1938), 23–24.

[256] Richard Bozarth, "Meaning of Evolution," *American Atheist* (Feb 1978), 19–30. For those that may find it interesting, in this article, he is referring to atheism as a religion.

[257] Dobzhansky, "Light of Evolution."

[258] "Robert Blatchford," http://en.wikipedia.org/wiki/Robert_Blatchford, 23 April 2011. He was a prominent atheist and opponent of eugenics. He was also an English patriot. In the early 1920s, after the death of his wife, he turned toward spiritualism.

[259] Ibid.

[260] Robert Blatchford, *God and My Neighbour* (Chicago: Clarion Press, Charles H. Kerr & Co., 1911), 158160.

[261] *Science and Creationism.*

[262] T. V. Wollaston, "The Annals of the Magazine of Natural History, including Zoology, Botany, and Geology," vol. 5, London, 1860, 132-143.

[263] *Science and Creationism.*

[264] Ibid.

[265] This includes not mentioning God as the deist God, gnostic God, Jesus, or Darwin's God.

[266] *Science and Creationism.*

[267] Ernst Mayr, *One Long Argument, Charles Darwin and the Genesis of Modern Evolutionary Thought* (Cambridge, MA: Harvard University Press, 1991), 99.

[268] King James Bible, http://www.kingjamesbibleonline.org/Genesis-Chapter-2/, 8 May 2014.

[269] Ibid.

[270] Charles Hodge, *What Is Darwinism?* (Princeton, NJ, New York: Scribner, Armstrong, And Company, 1874), 173.

[271] Ibid., 22.

[272] "Charles Darwin and the Early Evolutionists," http://spot.colorado.edu/~friedmaw/Early_Evolution/Homepage.html, 13 July 2009, Colorado University.

[273] Samuel Butler, *Evolution, Old & New*, 18-19.

[274] Ibid., 10, footnote 18.

[275] Mayr, *One Long Argument*, 90.

[276] Adrian Desmond, *Huxley, From Devil's Disciple to Evolution's High Priest* (Reading, MA: Addison-Wesley, 1994), 392.

[277] Ibid., 391.

[278] Ibid.

[279] Mayr, *One Long Argument*, 99.

[280] Desmond, *Huxley*, 392.

[281] Ibid.

[282] Mayr, *One Long Argument*, 91.

[283] Ibid., 99.

[284] A. Hallam, *Lyell's Views on Organic Progression, Evolution and Extinction* (London: Geological Society, Special Publications 1998), v. 143, 133–136, 135 (midway, right column), doi:10.1144/GSL.SP.1998.143.01.11.

[285] Mayr, *One Long Argument*, 99.

[286] Wallace, "Limits of Natural Selection."

[287] Mayr, *One Long Argument*, 99–100.

[288] Ibid.

[289] Browne, *Power of Place*, 247.

[290] Ibid.

[291] Ibid.

[292] Ibid.

[293] Ibid., 248.

[294] Ibid.

[295] Ibid.

[296] Ibid., 251.

[297] Hodge, *What Is Darwinism?* 82.

[298] Mayr, *One Long Argument*, 94.

[299] Ibid.

[300] Ibid.

[301] Hodge, *What Is Darwinism?* 23.

[302] Ibid., 23-24.

[303] Browne, *Power of Place*, 370.

[304] Ibid.

[305] Ibid.

[306] Ibid.

[307] Ibid.

[308] Herbert Spencer, *Principles of Biology*, Vol. I, Williams and Norgate, 14, Henrietta Street; Covent Garden, London; And 20, South Frederick Street, Edinburgh, 1864, 444.

[309] Ibid., 445.

[310] Darwin, *Origin of Species* (1872), 234 (last sentence).

[311] Wallace, *Darwinism*, 113.

[312] Spencer, Herbert—Social Darwinism," http://www.crf-usa.org/bria/bria19_2b.htm, 31 Mar 08.

[313] Wallace, *Darwinism*, 113.

[314] Daniel J. Kevles, *In the Name of Eugenics*, Alfred A. Knopf, New York, 1985, xiii.

[315] "Compulsory sterilization," http://en.wikipedia.org/wiki/Compulsory_sterilization, 5 Feb 2015.

[316] Russ Hodge, *Evolution: The History of Life on Earth*, (New York, Facts on File, Inc., Infobase Publishing, 2009), 89.

[317] Daniel J. Kevles, *In the Name of Eugenics*, Alfred A. Knopf, New York, 1985, 3.

[318] Ibid., 68.

[319] Ibid., 20. Darwin very favorably mentions Galton and his work at least eight times in *The Descent of Man*.

[320] Ibid., 54.

[321] Ibid., 100.

[322] Ibid., 111.

[323] Mein Kampf, Adolf Hitler, The Nationalist Socialist Movement, translated by James Murphy, 1924, Copyright © 2001-2010 Globusz' Publishing, 385.

[324] Ibid., 397.

[325] "Eugenics," http://en.wikipedia.org/wiki/Eugenics, February 12, 2015.

[326] Mein Kampf, Adolf Hitler, The Nationalist Socialist Movement, translated by James Murphy, 1924, Copyright © 2001-2010 Globusz' Publishing, 189.

[327] "Eugenics," http://en.wikipedia.org/wiki/Eugenics, February 12, 2015.

[328] Hitler, *Mein Kampf*, 189.

[329] Ibid., 189-190.

[330] Ibid., 192.

[331] Darwin, *Origin of Species* (1859), 79.

[332] Ibid.

[333] Darwin, *Descent of Man*, 35.

[334] Ibid., 10.

[335] Ibid., 201.

[336] Darwin, *Origin of Species* (1872), 171.

[337] Darwin, C. R. to Hooker, J. D., http://www.darwinproject.ac.uk/letter/entry-1924, 13 Jan 2015.

[338] "Was Darwin Right—Chance or Design?" http://www.wasdarwinright.com/doesitmatter.htm, 12 Jan 2011.

[339] Darwin, *Descent of Man*, 168.

[340] F. W. Nietzsche, *The Antichrist*, Translated from the German, with an introduction *by* H. L. Mencken, *New York*, Alfred A. Knopf, Copyright, 1918, By Alfred, A. Knopf, Inc. 2. Available from The Project Gutenberg EBook of The Antichrist, by F. W. Nietzsche, http://www.gutenberg.org/files/19322/19322h/19322-h.htm#THE_ANTICHRIST.

[341] "Edward Jenner," *Bartleby.com*, http://www.bartleby.com/people/Jenner-E.html, 3 March 2008.

[342] "Robert Koch," *Contagion: Historical Views of Diseases and Epidemics*, Harvard University Library Open Collections Program, http://ocp.hul.harvard.edu/contagion/koch.html, 1 Sep 2009.

[343] "Louis Pasteur (1822–95)," *Zephyrus*, http://www.zephyrus.co.uk/louispasteur.html, 19 Jan 2010

[344] Ibid.

[345] Ibid.

[346] Ibid.

[347] Ibid.

[348] Ibid.

[349] "Louis Pasteur (1822–95)," *Zephyrus*, http://www.zephyrus.co.uk/louispasteur.html, 19 Jan 2010.

[350] Ibid.

[351] Ibid.

[352] Ibid.

[353] Ibid.

[354] Ibid.

[355] "Pasteur Institute," http://en.wikipedia.org/wiki/Pasteur_Institute, December 21, 2013.

[356] Ibid.

[357] "Germ Theory of Disease," http://en.wikipedia.org/wiki/Germ_theory_of_disease, 20 Jan 2010.

[358] "Origin of Life—Spontaneous Generation," *AllAboutScience.org*, http://www.allaboutscience.org/origin-of-life.htm, 26 March 2010.

[359] "Louis Pasteur," *Panspermia.org*, http://www.panspermia.org/pasteur.htm, 8 Oct 08.

[360] "Louis Pasteur (1822–1895)," NHM Resource Center, http://www.accessexcellence.org/RC/AB/BC/Louis_Pasteur.php, 8 Oct 08.

[361] Browne, *Power of Place*, 310.

[362] "Gregor Mendel," *StrangeScience.net*, http://www.strangescience.net/mendel.htm, 2 Sep 08.

[363] "Gregor Mendel," *Wikipedia*; http://en.wikipedia.org/wiki/Gregor_Mendel; 2 Sep 08.

[364] *Evolutionary Psychology*, www.epjournal.net – 2007. 5(1): 52-69

[365] "Gregor Mendel, The Pea Plant Experiment," *Following the Path of Discovery*, http://www.juliantrubin.com/bigten/mendelexperiments.html, 2 Sep 08.

[366] Ibid.

[367] Ibid.

[368] Ibid.

[369] Ibid.

[370] Charles Darwin, *The Variation of Animals And Plants Under Domestication*, Vol. II, London, John Murray, Albemarle Street, 1868. See Chapter XXVII, Provisional Hypothesis of Pangenesis.

[371] Definition of Beijerinck, Martinus W, http://www.medterms.com/script/main/art.asp?articlekey=39058, 29 Jan 2015.

[372] Ibid.

[373] Science and Creationism.

[374] Evolutionary Psychology 5, no. 1: 52–69, Hiram Caton, Griffith University, Nathan 4111, Australia, hcaton2@bigpond.net.au, "Getting Our History Right," http://www.epjournal.net/articles/getting-ourhistory-right-six-errors-about-darwin-and-his-influence/, 15 January 2014.

[375] John Gribbin, *Science, A History 1543–2001* (New York: BCA, 2001), 533.

[376] "Archimedes," *Wikipedia*, http://en.wikipedia.org/wiki/Archimedes, 18 Nov 2012.

[377] For experiments with magnets and our surroundings, see http://my.execpc.com/~rhoadley/magmath.htm, 23 Dec 2013

[378] "Archimedes," *ThinkQuests*, http://library.thinkquest.org/25672/archimed.htm, 21 Feb 08.

[379] A seesaw is a long narrow board pivoted in the middle so as one end goes up the other goes down. http://en.wikipedia.org/wiki/Seesaw.

[380] "Lecture #9: You Can Get There from Here! Title: Spacecraft Trajectories," Wisconsin-Madison University, February 8, 1999 http://fti.neep.wisc.edu/~jfs/neep533.lecture9.trajectories.99.html, 19 July 2010.

[381] "Evolution as Fact and Theory," *Harvard Square Library*, http://www.harvardsquarelibrary.org/speakout/gould.html, 20 Oct 2011.

[382] "Comet Halley," http://csep10.phys.utk.edu/astr161/lect/comets/halley.html, 19 Feb 2011.

[383] "Halley's Comet," *Wikipedia*, http://en.wikipedia.org/wiki/Halley's_Comet, 5 Nov 2010.

[384] Ibid.

[385] "Mechanical Explanations of Gravitation," *Wikipedia*, http://en.wikipedia.org/wiki/Mechanical_explanations_of_gravitation, 21 May 2012.

[386] "Mendel's Genetics," http://anthro.palomar.edu/mendel/mendel_1.htm, 11 Aug 2010.

[387] "Anthrax Used by Koch and Pasteur to Prove Germ Theory of Disease in 19th C.," *Washington Times* (14 Oct 2001), http://www.mindfully.org/Health/Anthrax-Koch-Pasteur.htm, 11 Aug 2010.

388 Ibid.

389 In 1872, Pierre Pachet, professor of physiology at Toulouse, made the statement: "Louis Pasteur's theory of germs is ridiculous fiction."

390 Navin Sullivan, *Pioneer Germ Fighters,* (New York: Scholastic Book Services, 1962), 45.

391 Ibid.

392 Wallace, *Darwinism,* 36.

393 Nora Barlow, ed., "The Autobiography of Charles Darwin, 1809–1882," (St James's Place, London, Collins, 1958), 120.

394 Ibid., 90, 195.

395 Darwin, *Origin of Species* (1859), 447.

396 Ibid., 81.

397 Ibid.

398 Andrew Parker, *In the Blink of an Eye* (New York: Basic Books, 2003), 6.

399 William J. Bauer, PhD, "Review of Evolution of Living Organisms - By Pierre-Paul Grasse," http://www.icr.org/article/review-evolution-living-organisms-by-pierre-pual-g/, January 29, 2015.

400 Ibid.

401 Pierre-Paul Grasse, *Evolution of Living Organism* (New York: Academic Press, 1977), 103

402 Darwin, *Origin of Species* (1859), 81.

403 Ibid.

404 Ibid.

405 Ibid.

406 Ibid., 413.

407 Baron G. Cuvier, *Theory of the Earth,* 5th ed., William Blackwood, Edinburgh, and T. Cadell, Strand, London, 1827, 84.

408 Darwin, *Origin of Species* (1859), 189.

409 "Gordon's Introduction to Cells," http://www.earthlife.net/cells.html, 17 Oct 2010

410 "From One Genome, 200 Types of Cells," http://www.indianexpress.com/news/from-one-genome-200types-of-cells/429231/, 17 Oct 2010.

411 "How Evolution Became a Religion," *National Post* (2000), By Michael Ruse.

412 "The Silent Landscape: The Scientific Voyage of HMS *Challenger,*" http://findarticles.com/p/articles/mi_m1134/is_8_112/ai_108551832/, 25 Aug 09.

413 Ibid.

[414] Ibid.

[415] "The Voyage of the Challenger," SUNY, http://life.bio.sunysb.edu/marinebio/challenger.html, 24 Aug 09.

[416] Note that the term "evolutionists" used here by Mayr portrays "evolution" as an implied synonym for "natural selection."

[417] Ernst Mayr, *What Evolution Is* (New York: Basic Books, 2001), 281.

[418] Ibid.

[419] Parker, *Blink of an Eye*, 6.

[420] Ibid.

[421] Ibid.

[422] Weiner, *Beak of the Finch*, 188.

[423] Ernst Mayr, *What Evolution Is* (New York: Basic Books, 2001), 118.

[424] Ibid., 118.

[425] Parker, *Blink of an Eye*, 291.

[426] Ibid., 262.

[427] "Examples of the Evolutionary Arms Race among the Extinct Paleozoic Trilobites, Anecdotes in Trilobite Evolution," http://www.fossilmuseum.net/Evolution/TrilobiteArmsRace.htm, 15 January, 2014.

[428] Ibid.

[429] Ibid.

[430] "Trilobite," *Bing.com,* http://www.bing.com/Dictionary/search?q=define+trilobite&qpvt=trilobite+definition&FORM=DTPDI A, 9 Dec 2012.

[431] Richard Dawkins, "The Evolutionary Future of Man," *Economist* 328, (Sept.11, 1993), 87. Also published on, 6 Dec 2012.

[432] Examples of the evolutionary arms race among the extinct Palaeozoic trilobites, http://www.fossilmuseum.net/Evolution/TrilobiteArmsRace.htm, 13 April 2009.

[433] Wallace, *Darwinism*, 94–95.

[434] Ibid., 94–95.

[435] Ibid.

[436] Niles Eldredge and Ian Tattersall, *The Myths of Human Evolution* (New York: Columbia Univ., 1982), 40.

[437] Ibid.

[438] Ibid.

439 Ibid.

440 "Definition," *Webster*, http://www.merriam-webster.com/dictionary/personification, 20 June, 2012.

441 Ibid.

442 Darwin, *Origin of Species* (1859), 189.

443 Darwin, *Origin of Species* (1861), 84–85.

444 Ibid.

445 Ibid.

446 Ibid.

447 Ibid., 85.

448 Ibid., 80.

449 Ibid., 83.

450 Ibid., 5.

451 "Letter to Asa Gray, September 5th, 1857," by Charles Darwin, http://www.stephenjaygould.org/library/darwin_gray.html, 9 July 2012.

452 Darwin, *Origin of Species* (1859), 269.

453 Ibid.

454 Ibid., 83.

455 Ibid., 62.

456 Ibid., 459.

457 Robert Young, *Darwin's Metaphor, Nature's Place in Victorian Culture*, Cambridge University Press; 1ST edition (October 31, 1985), 119.

458 "Metaphor," *Wikipedia*, http://en.wikipedia.org/wiki/Metaphor, 26 April 2008.

459 Ibid.

460 Miracle, Merriam Webster http://www.merriam-webster.com/dictionary/miracle, January 15, 2015

461 "The Swerve: How the World Became Modern," http://en.wikipedia.org/wiki/The_Swerve:_How_the_World_Became_Modern, 22 April 204 463 Stephen Greenblatt, "The Answer Man," *The New Yorker* (New York: Aug 8, 2011), 28.

462 Ibid., 30.

463 Miracles are extensively tested by many different bodies in the Catholic Church over long periods of time before they are accepted.

464 "Miracle," *Wikipedia*, en.wikipedia.org/wiki/Miracle, 15 Jan 2014.

[465] Miracles, Stanford Encyclopedia of Philosophy, http://stanford.library.usyd.edu.au/archives/sum1999/entries/miracles/, 14 March 2015

[466] The tests are not the Newton-like "foretelling tests" but rather "testimonial tests" like those of natural selection and miracles.

[467] Randall Sullivan, *The Miracle Detective* (New York: Grove Press, 2004), 26–27.

[468] Ibid., 27.

[469] Ibid., 27–28.

[470] "Sir Charles Lyell on Geological Climates and the Origin of Species (S146: 1869)," http://people.wku.edu/charles.smith/wallace/S146.htm, 17 Oct 2011.

[471] Ibid.

[472] Ibid.

[473] Wallace, "The Limits of Natural Selection as Applied to Man (S165: 1869/1870)," http://www.wku.edu/~smithch/wallace/S165.htm, 9 Sep 09.

[474] Karen James, "A guest post by Wallace's Rottweiler on the 150th anniversary of natural selection," *The HMS Beagle Project* (2014), http://blog.hmsbeagleproject.org/.

[475] *The Alfred Russel Wallace Website*, http://wallacefund.info/, 3 Sep 2014.

[476] Science and Creationism.

[477] Wallace, "Limits of Natural Selection."

[478] Ibid.

[479] Ibid.

[480] Ibid.

[481] Ibid.

[482] Ibid.

[483] Ibid.

[484] Ibid.

[485] Ibid.

[486] Ibid.

[487] Ibid.

[488] Ibid.

[489] Butler's "intellectual violence" may be considered ideas that are in opposition to the worldview that is being used.

[490] Butler, *Evolution*, 19.

[491] Ibid., 20.

[492] William Paley, *Natural Theology: or, Evidences of the existence and attributes of the Deity, collected*, (New York, Sheldon & Company, 8 Murray Street, 1879), 293.

[493] Darwin, *Origin of Species* (1859), 131.

[494] Ibid., 198.

[495] Ibid., 81.

[496] "Worldview," *Wikipedia*, http://en.wikipedia.org/wiki/Worldview, 10 Jan 2014.

[497] "Michael Faraday," http://chem.ch.huji.ac.il/history/faraday.htm, 21 April 2009.

[498] Ibid.

[499] Lovejoy, Arthur O. Lovejoy: *The Great Chain of Being: A Study of the History of an Idea*, Cambridge, Massachusetts: Harvard University Press (1936:7)

[500] "The Mechanical Philosophers," http://history.wisc.edu/sommerville/351/351-192.htm, 5 Nov 2010.

[501] Ibid.

[502] Albert Einstein and Leopold Infeld, *The Evolution of Physics* (New York: Simon & Schuster, 1961 [Originally published in 1938]), 151.

[503] "Origin of Life—Spontaneous Generation," http://www.allaboutscience.org/origin-of-life.htm, 28 Dec 2013.

[504] Ibid.

[505] "Spontaneous Generation," *InfoPlease.com*, http://www.infoplease.com/cig/biology/spontaneousgeneration.html, 3 Feb 09.

[506] Ibid.

[507] Ibid.

[508] Darwin, *Origin of Species* (1860), 490

[509] Darwin, *Origin of Species* (1859), 490.

[510] Letter 4065, Darwin, C. R. to Hooker, J. D., 29 Mar 1863. See http://www.darwinproject.ac.uk/entry-4065 accessed on Tue Feb 17 2015.

[511] Juli Peretó, Jeffrey L. Bada, and Antonio Lazcano, "Charles Darwin and the Origin of Life", http://www.ncbi.nlm.nih.gov/pmc/articles/PMC2745620/, 31 Mar 2014

[512] "The Life and Times of Louis Pasteur," http://louisville.edu/library/ekstrom/special/pasteur/cohn.html, 3 Feb 09, 19 April 2012

[513] "The Life and Times of Louis Pasteur," keynote address by David V. Cohn School of Dentistry, University of Louisville, Feb. 11, 1996, for the centennial celebration of the death of Pasteur that was sponsored jointly in 1996 at the University of Louisville by the University,

the Pasteur Institute of Paris, and the Alliance Française de Louisville. It has been revised for presentation on this website: http://pyramid.spd.louisville.edu/~eri/fos/interest1.html; 10 Dec 2013

[514] Wallace, "Limits of Natural Selection."

[515] Mayr, *One Long Argument,* 99.

[516] "The Newtonian Worldview," *Principia,* http://pespmc1.vub.ac.be/NEWTONWV.html, 15 Nov 2010.

[517] Browne, *Power of Place,* 304.

[518] Charles Darwin, *The Expression of the Emotions in Man and Animals,* (New York, D. Appleton And Company, 1899).

[519] "Biography of Sir Charles Bell," http://www.whonamedit.com/doctor.cfm/2103.html, 3 May 08.

[520] Ibid.

[521] Science and Creationism.

[522] Ibid.

[523] Ibid.

[524] R. A. Fisher, Sc.D., F.R.S., *The Genetical Theory of Natural Selection* (Oxford, UK: Clarendon Press, 1930), vii.

INDEX

INDEX

www.ingramcontent.com/pod-product-compliance
Lightning Source LLC
Chambersburg PA
CBHW031822170526
45157CB00001B/147